Introduction to Differential Equations and Linear Algebra

Fourth Edition

Alan Parks
Lawrence University,
Appleton, Wisconsin

♠ This project puts text material, carefully coordinated with lectures and homework, into the hands of students at cost. It was begun during the summer of 2001.

Contents

Introduction

> The senses represent objects only which exist externally; and sensible ideas all refer to them. But of these sensible ideas the soul forms to itself a variety of other ideas ... but which no longer represent objects really existing. ... Here the soul exerts a new facility called the power of abstraction.
>
> Leonhard Euler, [**8**, p.330]

A *differential equation*, abbreviated *DE*, involves the derivatives of a function; they arise in the study of almost every physical law in the sciences. In the DE $dx/dt = t^2$, the differential notation indicates that x is a function of t, and a *solution* to the equation is a particular function $x = f(t)$ that makes the equation true.[1] This particular equation is easy to solve by integration, and we see that

$$(0.1) \qquad x = \int t^2 \cdot dt = \frac{t^3}{3} + c \quad \text{where } c \text{ is an arbitrary constant}$$

The infinitely many solutions (coming from infinitely many c) are distinguished from each other by their value when $t = 0$. We typically write $x(0)$ or x_0 for the value of x when $t = 0$, and observe from (0.1) that $x_0 = c$. Thus, if we specify the value of x_0, we arrive at a unique solution x. For instance, if we say $x_0 = -2$, then we have $x = t^3/3 - 2$ as the unique solution. The problem consisting of the equation and the initial value is called an *initial value problem*, abbreviated *IVP*. Most DE's that occur in practice are IVP's.

This course studies the most basic and common IVP's, with particular attention to a wide variety of application problems. Our point of view is *mathematical* – in other words, we are interested in how mathematical features of application problems lead to mathematical information about those

---

[1]The variable t will be used almost universally as the independent variable in this course. This is customary; it probably stems from the fact that *time* is very often the independent variable in a physical model.

problems. As persons of a pronounced, general curiosity, we would like to have some understanding of the applications on their own terms – as they arise in their own disciplines, and the discussion in this text and in class will take steps in that direction, but our emphasis will be on the reasoning that produces mathematical solutions and mathematical information about those solutions.

The other main topic of the course is *linear algebra*: the study of the large body of *linear problems*. These problems, and the computational algebra involved in solving them, are basic to almost all of mathematics and especially to application problems. Once you have learned this material, you will see that linear problems are *easy* in the sense that we can solve them by applying a fairly short list of facts and techniques.

In other words, both subjects – DE's and linear algebra – are prevalent in a variety of settings. Moreover, they intertwine in some very interesting ways, and so it is extremely profitable to study them together. For instance, the terminology and methods of linear algebra show up over and over again in the solution of DE's. Linear algebra will, in fact, furnish insights into DE's.

In a first course, we cannot be comprehensive with respect to DE's, or to linear algebra. We repeat what we said above: this course contacts the most basic and common problems. We are especially interested in developing the ability to recognize common abstract forms.

Much of your work in the course will have the computational flavor of calculus – but there will be more emphasis on the reasoning that justifies computation. You may well need to review some of the calculus; keep your calculus book handy. Speaking of books, textbooks introducing differential equations and/or linear algebra are legion. The bibliography on p.227 contains a list of some standard texts: [**3**], [**4**] introduce linear algebra, [**1**], and [**2**] DE's. Feel free to use these references for additional problems and for alternative points of view.

Problems are listed at the end of each chapter; many of these will be assigned, some will be done in class. The problems have been chosen carefully, and you are encouraged to think about all of them.

CHAPTER 1

Preliminaries

We denote by $\mathbb{R}$ the set of real numbers. For a positive integer n, we write $\mathbb{R}^n$ for n-space: an element v of $\mathbb{R}^n$ has n real number *coordinates* $v[1], v[2], \ldots, v[n]$, and we call v a *vector*. At this level of mathematics, we drop the distinction between *points* and *vectors*, since both are defined by the number of their coordinates (sometimes called *components*). In fact, we will gradually generalize the use of the word *vector* in other ways as we go along – that generalization is one of the central ideas of linear algebra.

1. Integration and the Semi-Definite Integral

It will come as no surprise that equations involving the derivative will often be solved by integration. So, some review presents itself!

The most important idea is this: a continuous function on an interval necessarily has an antiderivative there. In many typical cases we know how to find a formula for the antiderivative, but the *theoretical* existence of the antiderivative is guaranteed whether we can find a particular formula for it or not. This fact will serve several purposes in the course.

As to particular antiderivative formulas, we give a table of those most commonly used. You should be able to use most of these without looking them up, but we will feel free to use this table whenever we need to.

We have put the table on its own page so it will be easy to reference.

$$\int x^n dx = \frac{x^{n+1}}{n+1} \quad \text{if } n \neq -1$$

$$\int \frac{dx}{x} = \ln|x|$$

$$\int \cos(kx)dx = \frac{1}{k}\sin(kx)$$

$$\int \sin(kx)dx = -\frac{1}{k}\cos(kx)$$

$$\int e^{kx}dx = \frac{1}{k}e^{kx}$$

$$\int \tan(x)dx = \ln|\sec(x)|$$

$$\int \sec(x)dx = \ln|\sec(x) + \tan(x)|$$

$$\int \cos^2(x)dx = \frac{x}{2} + \frac{\sin(2x)}{4}$$

$$\int \frac{dx}{x^2+1} = \arctan(x)$$

$$\int \frac{dx}{\sqrt{1-x^2}} = \arcsin(x)$$

$$\int \frac{dx}{(x-a)(x-b)} = \frac{1}{a-b}\ln\left|\frac{x-a}{x-b}\right|$$

$$\int \sqrt{1-x^2}\cdot dx = \frac{1}{2}\left(x\cdot\sqrt{1-x^2} + \arcsin(x)\right)$$

We also have

$$\int \exp(a\cdot t)\cdot\cos(b\cdot t)\cdot dt = \frac{\exp(a\cdot t)}{a^2+b^2}\cdot\left[a\cdot\cos(b\cdot t) + b\cdot\sin(b\cdot t)\right]$$

$$\int \exp(a\cdot t)\cdot\sin(b\cdot t)\cdot dt = \frac{\exp(a\cdot t)}{a^2+b^2}\cdot\left[-b\cdot\cos(b\cdot t) + a\cdot\sin(b\cdot t)\right]$$

Parts. $\int u\cdot dv = u\cdot v - \int v\cdot du.$

$$\int x^n\cdot\left\{e^{kx} \text{ or } \cos(kx) \text{ or } \sin(kx)\right\}dx \qquad \text{uses} \qquad u = x^n$$

$$\int x^n\cdot\left\{\ln(x) \text{ or arcsomething }\right\}dx \qquad \text{uses} \qquad dv = x^n dx$$

We introduce a useful notation. The definition is abstract and will be best understood via a few examples. But here is the definition: let $f(t)$ be a continuous function and let α be a real number. Then

$$\int_\alpha f(t) \cdot dt$$

denotes the antiderivative of $f(t)$ that has value 0 when $t = \alpha$. To say it again, if we write

$$G(t) = \int_\alpha f(t) \cdot dt$$

then we have

$$G'(t) = f(t) \quad \text{and} \quad G(\alpha) = 0$$

The function $G(t)$ is called the *semi-definite integral* of $f(t)$ at $t = \alpha$.

Examples will show us how to find the semi-definite integral. Watch:

$$\int_1 (3 \cdot t^2 + 2) \cdot dt = t^3 + 2 \cdot t \Big|_1 = (t^3 + 2 \cdot t) - (1^3 + 2 \cdot 1) = t^3 + 2t - 3$$

Notice that the derivative of $t^3 + 2t - 3$ is $3t^2 + 2$ and notice that $t^3 + 2t - 3$ is 0 when $t = 1$.

Another example:

$$\int_0 \sin(t) \cdot dt = -\cos(t) \Big|_0 = -\cos(t) + \cos(0) = 1 - \cos(t)$$

You should check that $1 - \cos(t)$ meets the requirements of the semi-definite integral of $\sin(t)$ at $t = 0$.

The examples show how to compute the semi-definite integral when we know an antiderivative formula. In general, if $F(t) = \int f(t) \cdot dt$ is an arbitrary antiderivative of $f(t)$, then the semi-definite integral at $t = \alpha$ is this:

$$\int_\alpha f(t) \cdot dt = F(t) - F(\alpha)$$

Thus, we evaluate the antiderivative only at the lower limit of integration – that's why it's the *semi*-definite integral.

2. The Complex Exponential Function

The *complex exponential function* will play a very prominent role in the course. In class we will talk about the several ways this function can be defined; we have chosen the definition that most suits the purposes of this course.

For the (ordinary) exponential function from calculus, we will use the common notation $\exp(x)$ for e^x whenever it seems easier to read. The solution of many, many DE's will depend on the following very simple formula: for each real number k, we have

$$\frac{d}{dt} e^{k \cdot t} = k \cdot e^{k \cdot t}$$

A matter of terminology. If x is a function of time t, then you know that x' is the *rate of change* of x. The ratio x'/x is the *relative rate* of x. When $x = e^{kt}$, we have $x' = k \cdot e^{kt} = k \cdot x$, and so the relative rate is $x'/x = k$. To repeat: the exponential constant k is the *relative rate* of e^{kt}.

We want to allow k in $\exp(k \cdot t)$ to be an arbitrary *complex number*.[1] The variable t will remain real. Let's move toward the needed formula.

One of the most important properties of the exponential function is this:

$$\exp(a + b) = \exp(a) \cdot \exp(b)$$

We will want this identity to hold when complex numbers are involved. With this in mind, consider the expression $\exp(a + i \cdot b)$ when a, b are real numbers and $i^2 = -1$. Using the identity just mentioned, we would need to have

$$\exp(a + i \cdot b) = \exp(a) \cdot \exp(i \cdot b)$$

The expression $\exp(a) = e^a$, as before. To define the exponential of a complex number, we apparently need to define $\exp(i \cdot b)$ for each real number b. The

---

[1]Do we really *need* complex numbers? We will see that they arise inevitably even in problems that involve purely real values!

formula we want was discovered by Euler.[2] Here goes.

(1.1) $\exp(i \cdot b) = \cos(b) + i \cdot \sin(b)$ for all real numbers b

This is an extremely important and suggestive definition. To get started with it, notice that when $b = 0$ we have

$$\exp(i \cdot 0) = \cos(0) + i \cdot \sin(0) = 1$$

This makes sense, since we seem to be saying that $e^{i \cdot 0} = e^0 = 1$. We will see that $\exp(i \cdot b)$ has properties very similar to those of the calculus exponential function. Here are some of those properties; the proofs in each case will be a direct calculation from the definition, and they will be done in class.

PROPOSITION 1.1. *Let r, s be real numbers. Then*

(a) We have $\exp(i \cdot (r + s)) = \exp(i \cdot r) \cdot \exp(i \cdot s)$.
(b) For each integer q, we have $(\exp(i \cdot r))^q = \exp(i \cdot q \cdot r)$.

Equation (1.1) allows us to define $\exp(k)$ where k is an arbitrary complex number. Let $k = a + i \cdot b$, and we define

(1.2) $\exp(a + i \cdot b) = e^a \cdot \left(\cos(b) + i \cdot \sin(b) \right)$

This definition allows us to consider the function $\exp(k \cdot t)$ where k is a complex number and t is a real variable. Here are the properties of this function. Notice that when k is a *real* number, $\exp(k \cdot t)$ is as it was in calculus, since the cosine/sine part is 1. The following will be proved in class.

PROPOSITION 1.2. *We have*

(a) If p, q are complex numbers then $\exp((p + q) \cdot t) = \exp(p \cdot t) \cdot \exp(q \cdot t)$
(b) If k is a complex number, then $[\exp(k \cdot t)]' = k \cdot \exp(k \cdot t)$.

---

[2]See [**5**, Chapters VII, VIII]. There are many ways to derive the formula; it can be done via power series, from the many trigonometric identities, and in other ways.

There are a couple more identities that are quite useful. Let c be a real number, and compute

$$\exp(-i \cdot c) = \cos(-c) + i \cdot \sin(-c)$$

The cosine function is *even*: $\cos(-c) = \cos(c)$; the sine function is *odd*: $\sin(-c) = -\sin(c)$. In view of this, we see that

(1.3) $$\exp(-i \cdot c) = \cos(c) - i \cdot \sin(c)$$

Putting the formula (1.3) together with $\exp(i \cdot c)$ we see that

$$\exp(i \cdot c) = \cos(c) + i \cdot \sin(c)$$
$$\exp(-i \cdot c) = \cos(c) - i \cdot \sin(c)$$

Adding these equations:

$$\exp(i \cdot c) + \exp(-i \cdot c) = 2 \cdot \cos(c)$$

so that

(1.4) $$\cos(c) = \frac{\exp(i \cdot c) + \exp(-i \cdot c)}{2}$$

Similarly,

(1.5) $$\sin(c) = \frac{\exp(i \cdot c) - \exp(-i \cdot c)}{2i}$$

These identities allow us to convert cosine and sine to the exponential function; we will make use of this in the next section.

3. Cosine/Sine Form and Cosine Form

For application problems, it will be useful to have a couple of different ways to express combinations of complex exponential functions. First, we note that the definition (1.1) allows us to convert the complex exponential into cosines and sines. In many cases, this rids the expression of non-real numbers. Here's a typical example. (Don't forget (1.3)!)

$$(3 + 2 \cdot i) \cdot \exp(4 \cdot i \cdot t) + (3 - 2 \cdot i) \cdot \exp(-4 \cdot i \cdot t)$$
$$= (3 + 2 \cdot i) \cdot (\cos(4t) + i \cdot \sin(4t))$$
$$+ (3 - 2 \cdot i) \cdot (\cos(4t) - i \cdot \sin(4t))$$

This simplifies when we put the cosines together and the sines together.

$$= (3 + 2i + 3 - 2i) \cdot \cos(4t) + ((3 + 2i)i - (3 - 2i)i) \cdot \sin(4t)$$
$$= 6 \cdot \cos(4t) + (3i - 2 - 3i - 2) \cdot \sin(4t)$$
$$= 6 \cdot \cos(4t) - 4 \cdot \sin(4t)$$

The resulting form is called *cosine/sine form* – for obvious reasons. This form is sometimes preferred in applied work. The point is that equations (1.1) and (1.3) allow us to convert the exponential to cosines and sines.

We remind you of standard terminology: for a constant k, the functions $\cos(k \cdot t)$ and $\sin(k \cdot t)$ have *frequency* $k/(2\pi)$ and *period* $2\pi/k$.

Here is a more complicated example. We pull off the real-exponential e^{2t} and then use the identities for cosine and sine.

$$(-5 + 3i) \cdot \exp((2 + i)t) + (-5 - 3i) \cdot \exp((2 - i)t)$$
$$= (-5 + 3i) \cdot e^{2t} \cdot \exp(it) + (-5 - 3i) \cdot e^{2t} \cdot \exp(-it)$$
$$= e^{2t} \cdot \left[(-5 + 3i) \cdot \big(\cos(t) + i\sin(t)\big) + (-5 - 3i) \cdot \big(\cos(t) - i\sin(t)\big)\right]$$
$$= e^{2t} \cdot \left[(-5 + 3i - 5 - 3i) \cdot \cos(t) + (-5i - 3 + 5i - 3) \cdot \sin(t)\right]$$
$$= e^{2t} \cdot \left[-10 \cdot \cos(t) - 6 \cdot \sin(t)\right]$$

You may have noticed the prevalence of complex conjugates in the two cosine/sine calculations – such conjugates typically occur in solutions to DE's, as we will see in Chapter 7.

There is another form useful for application problems: the *cosine form*

$$A \cdot \cos(wt + q) \quad \text{where} \quad A, w, q \quad \text{are constants}$$

The number A is the *amplitude*, since the values of the expression $A \cdot \cos(wt+q)$ oscillate between $\pm A$. The number q is the *phase*. Let's see how to convert cosine/sine form to cosine form.

Problem.   Convert $3 \cdot \cos(2t) - 4 \cdot \sin(2t)$ to cosine form.

Solution.   The trick is to write the coefficients $(3, -4)$ in polar coordinates. The distance to the origin (the "r" of polar) is $\sqrt{3^2 + 4^2} = 5$. (Nice numbers!) There is an angle θ such that $3 = 5 \cdot \cos(\theta)$ and $-4 = 5 \cdot \sin(\theta)$. Actually, $\theta = \arcsin(-4/5)$ will work, but we will keep writing θ for the angle. Then

$$3 \cdot \cos(2t) - 4 \cdot \sin(2t) = 5 \cdot \cos(\theta) \cdot \cos(2t) + 5 \cdot \sin(\theta) \cdot \sin(2t)$$

The cosine addition formula kicks in.

$$5 \cdot \cos(\theta) \cdot \cos(2t) + 5 \cdot \sin(\theta) \cdot \sin(2t) = 5 \cdot \cos(2t - \theta)$$

and so $3 \cdot \cos(2t) - 4 \cdot \sin(2t) = 5 \cdot \cos(2t - \theta)$, and we have our cosine form.

This example leads to an important general observation.

Amplitude from cosine/sine form.   The amplitude of

$$A \cdot \cos(wt) + B \cdot \sin(wt) \quad \text{where} \quad A, B, w \quad \text{are constants}$$

is $\sqrt{A^2 + B^2}$.

The cosine addition formula can also be used to go from cosine form to cosine/sine form (this is rarely used).

$$3 \cdot \cos(5t + \pi/3) = 3 \cdot \cos(5t) \cdot \cos(\pi/3) - 3 \cdot \sin(5t) \cdot \sin(\pi/3)$$

$$= \frac{3}{2} \cdot \cos(5t) - \frac{3\sqrt{3}}{2} \cdot \sin(5t)$$

4. Poly-Exponential Functions

There is an important class of functions that comes up in the solution of several types of DE's. These functions do not have an official name, and so, since they involve polynomials and the exponential function, we have decided to call them the *poly-exponential* functions. To begin to describe these functions, we say that $\exp(\alpha \cdot t)$ is poly-exponential for every complex number α. Also, every polynomial function is poly-exponential. Furthermore, sums and products of poly-exponential functions are poly-exponential. Here is what this amounts to: if we are given complex numbers $\beta_1, \ldots, \beta_k$ and polynomials $P_1(t), \ldots, P_k(t)$, then the function

$$(1.6) \qquad f(t) = \sum_{j=1}^{k} P_j(t) \cdot \exp(\beta_j \cdot t)$$

is poly-exponential, and every poly-exponential function has this form. As special cases, we can get a single polynomial $P(t) = P(t) \cdot \exp(0 \cdot t)$, and we can get a single exponential $\exp(a \cdot t) = 1 \cdot \exp(a \cdot t)$.

Equations (1.4) and (1.5) show that $\cos(\alpha \cdot t)$ and $\sin(\alpha \cdot t)$ are poly-exponential for every real number α.

Later in the course, we will need the following fact.

PROPOSITION 1.3. *Let $f(t)$ be an antiderivative of a poly-exponential function. Then $f(t)$ is poly-exponential.*

PROOF. Writing $f'(t)$ as in (1.6), we obtain $f(t)$ by integrating. The antiderivative of a polynomial is a polynomial, and integration by parts shows that the antiderivative of $p(t) \cdot e^{at}$, where $p(t)$ is a polynomial and $a \neq 0$, is poly-exponential as well. This does it. $\square$

CHAPTER 2

First Order Differential Equations

A *first order* DE is an equation in which only the first derivative of the unknown function is involved. Here is arguably the most common such equation; let k be a constant:

$$\frac{dy}{dt} = k \cdot y$$

The equation, which says that the relative rate y'/y is constant, occurs abundantly.[1] Writing y_0 for $y(0)$, Proposition 1.2 in Chapter 1 shows that $y = y_0 \cdot e^{kt}$ is a solution to this equation and that k can be an arbitrary complex number constant. Later we will see that this is the *only* solution.

The rest of this chapter concerns itself with the two types of first order DE's that can be solved routinely.

1. Separable Differential Equations

There are several technicalities here, but the main idea is very simple. Let's start with an example that illustrates this idea without worrying about technicalities. An object strikes the surface of a viscous fluid at 10 feet per second. As the object travels through the fluid, we will assume that its velocity decreases at a rate proportional to the square of the velocity. Writing v for the

---

[1]Occurrences when $k > 0$: the DE is the *first order reaction equation* of chemistry, the *Malthusian Law* of population growth in biology, the equation describing nuclear chain reactions in physics, the growth of an investment at continuously compounded interest rate, and so on. When $k < 0$, the equation describes radioactive decay, reaction to simple types of resistance or friction, the concentration of a substance in the body that is gradually eliminated, the *mortality equation* of population decrease, and so on.

velocity, we are saying that $v_0 = 10$ and that $v' = -k \cdot v^2$ for some constant k when $t \geq 0$. In the DE

$$\frac{dv}{dt} = -k \cdot v^2$$

we can separate the v and dv from the dt:

$$\frac{dv}{v^2} = -k \cdot dt$$

This form invites integration. Because $v = 10$ when $t = 0$, we use semi-definite integrals to reflect this:

$$\int_{10} \frac{dv}{v^2} = \int_0 (-k) \cdot dt \quad \text{which is} \quad \frac{1}{10} - \frac{1}{v} = -k \cdot t$$

We can solve this last equation for v, but let's focus on the main idea: we *separated* v and t multiplicatively and integrated.

Here's what a separable DE looks like in general, this time using y as the unknown function:

(2.1) $$\frac{dy}{dt} = f(y) \cdot g(t)$$

An IVP would give y_0. The idea of solution is to *separate* y from t:

$$\frac{dy}{f(y)} = g(t) \cdot dt$$

and then we integrate.

Let's do another example. Later we will see that the equation involved comes from an application; for now it's just a good example. The number k is a constant in this problem.

(2.2) $$\frac{dy}{dt} = k \cdot (10 - y) \quad \text{and} \quad y_0 = 10$$

We can try separation

$$\frac{dy}{10 - y} = k \cdot dt$$

but we run into a problem with the initial value. When $t = 0$, the fact that $y = 10$ tells us that the expression $10 - y$ is 0. In that case we can't divide by $10 - y$ and so the separation is invalid.

Here's what the previous problem looks like in general: we have equation (2.1) with initial condition y_0 where $f(y_0) = 0$, so that we can't divide by $f(y)$. This problem is actually a remarkably fortunate situation! The *constant function* $y = y_0$ is a *solution* to (2.1), for the left side dy/dt is 0 when we differentiate the constant $y = y_0$, while the right side is $f(y_0) \cdot g(t) = 0 \cdot g(t) = 0$, as well.

Constant solutions to DE's are often very important; a constant solution is called an *equilibrium*. We see that $y = 10$ is an equilibrium solution to the DE (2.2).

Let's change the initial value in (2.2), so that the separation of variables technique applies; say $y_0 = 5$ and then

$$\int_5 \frac{dy}{10 - y} = \int_0 k \cdot dt$$

makes sense as long as $y \neq 10$. We integrate.

$$\int_5 \frac{dy}{10 - y} = \int_0 k \cdot dt$$
$$- \ln|10 - y| + \ln|10 - 5| = k \cdot t - 0$$
$$- \ln|10 - y| + \ln 5 = k \cdot t$$

It is often useful in applied problems to resolve the absolute value on a logarithm. The appearance of $10 - y$ in the denominator tells us that $y \neq 10$. Since $y_0 = 5$ is less than 10, and since y is continuous (because it has a derivative), we see that y stays less than 10. Therefore, $|10 - y| = 10 - y$, and our equation is

$$\ln(5) - \ln(10 - y) = k \cdot t$$

We can take this equation further, and we'll do that in class. Let's pause to summarize what we've done, in general.

Solving a Separable DE   Given the IVP

$$(2.3) \qquad \frac{dy}{dt} = f(y) \cdot g(t) \quad \text{with} \quad y_0$$

if $f(y_0) = 0$, then $y = y_0$ is an equilibrium solution. If $f(y_0) \neq 0$, then we compute

$$\int_{y_0} \frac{dy}{f(y)} = \int_0 g(t) \cdot dt$$

to find a solution. If y_0 is not given, we need to consider both cases. ∎

There is an additional snag that can come up. Consider the equation

$$\frac{dy}{dt} = \frac{6 \cdot t^5 + 1}{5 \cdot y^4 + 1} \quad \text{and} \quad y_0 = 1$$

and you should be able to show that

$$(2.4) \qquad y^5 + y + 1 = t^6 + t$$

Here's the snag: there is no simple way to solve for y in this equation. In other words, there's no simple way to write down a solution $y = h(t)$ to this equation. It is customary to allow the equation (2.4) to serve as a solution; it is called a *solution equation* (or a *solution curve* since the equation seems to define a curve in the t, y-plane).

   Here is a more technical discussion of the non-equilibrium case. We include it to be complete, and you can skip this paragraph if you wish. Suppose that f, g are continuous in the equation (2.3), and suppose that y_0 is chosen with $f(y_0) \neq 0$. There is an open interval around y_0 in which $f(y)$ is not zero; in that interval define $F(y) = \int_{y_0} dy / f(y)$. We will use the variable z for $F(y)$, as well. Since $z = F(y)$ has a non-zero derivative, there is an inverse function

$F^{-1}(z) = y$ for the values of y in the open interval.[2] Then we have

$$\left[F^{-1}(z)\right]' = \frac{dy}{dz} = \frac{1}{dz/dy} = \frac{1}{1/f(y)} = f(y)$$

Keeping this in mind, we turn to $g(t)$. That function has an antiderivative $G(t) = \int_0 g(t) \cdot dt$. We claim that $y = F^{-1}(G(t))$ is a solution to the IVP. Indeed,

$$\frac{dy}{dt} = \left[F^{-1}\right]'(G(t)) \cdot G'(t) = f(y) \cdot g(t)$$

as needed. The equation $y = F^{-1}(G(t))$ can be written $F(y) = G(t)$, and that equation is the solution equation.

2. Examples and Applications

Whenever two variables, say x and y, are functions of a third variable t, there is a natural relationship between the derivatives dy/dx and dy/dt and dx/dt that comes from the Chain Rule.[3] If $dx/dt \neq 0$, then we have

(2.5)
$$\frac{dy}{dx} = \frac{dy/dt}{dx/dt}$$

There are situations in which the resulting equation in dy/dx is separable.

The Most Famous DE.   This is our affectionate name for a very commonly occurring equation that describes oscillations, vibrations, and other periodic behavior. Here it is.

(2.6)
$$\frac{dx}{dt} = -y \quad \text{and} \quad \frac{dy}{dt} = x$$

You might observe that this is the pattern of the derivatives of cosine and sine:

$$\cos'(t) = -\sin(t) \quad \text{and} \quad \sin'(t) = \cos(t)$$

---

[2]We are referring to a version of the *Inverse Function Theorem* of advanced calculus; we will not give a proof in this course. It may be difficult to find the domain in t of the solution y.

[3]The technicalities are somewhat complicated; we are again using the Inverse Function Theorem.

In fact, there are a host of solutions to this DE, all related to the trigonometric functions. In our present context, we use (2.5)

$$\frac{dy}{dx} = \frac{x}{-y} \quad \text{so that} \quad -y \cdot dy = x \cdot dx$$

Integrating, we find that

$$-\frac{y^2}{2} + \frac{y_0^2}{2} = \frac{x^2}{2} - \frac{x_0^2}{2} \quad \text{so that} \quad x_0^2 + y_0^2 = x^2 + y^2$$

The number $x_0^2 + y_0^2$ is a non-negative constant. Thus, the equation describes a circle centered at $(0,0)$ of radius $\sqrt{x_0^2 + y_0^2}$. You know that the points

$$(x, y) = (\cos(t), \sin(t))$$

lie on the unit circle $x^2 + y^2 = 1$. You can show that if r is a constant, then $x = r \cdot \cos(t)$, $y = r \cdot \sin(t)$ is a solution to the original DE. What circle does (x, y) lie on in this case?

Notice that the circles in x, y do not mention t, and so they do not give a *solution* to the original DE (2.6). However, they give curves in the x, y plane on which the solutions have to lie. These curves are called *solution equations* – similar to our use of this term previously, the equations in x, y seem to contain the graph of a solution. ∎

A Predator-Prey Model. We seek for a simple model of a population of x rabbits and y wolves, varying over time t, and reflecting two facts: rabbits breed and wolves eat rabbits. Here is an overview that will be discussed in class: The population of rabbits subsists on the land, and so, left to itself, it grows at a rate proportional to the population. The effect of the predator is that the rabbit population is damped at a rate jointly proportional to the number of rabbits and the number of wolves. The wolves depend on the rabbits, and so the wolf population increases at a rate jointly proportional to the number of wolves and rabbits, otherwise decreasing at a rate proportional to the number of wolves.[4] Here is our model, involving positive constants p, q, r, s. Remember that x is the population of rabbits, and y of wolves.

$$\frac{dx}{dt} = p \cdot x - q \cdot x \cdot y, \quad \frac{dy}{dt} = r \cdot x \cdot y - s \cdot y$$

This equation is called the *Lotka-Volterra equation*.[5] We will find the equilibria, and we will describe solution equations, assuming that x, y are positive.

As before, equilibria are constant solutions. If x, y are constant, then $dy/dt = 0 = dx/dt$, and we see that

$$p \cdot x - q \cdot x \cdot y = 0 \quad \text{and} \quad r \cdot x \cdot y - s \cdot y = 0$$

You should be able to do the algebra necessary to show that there are two equilibria: (1) $x = 0 = y$ and (2) $x = s/r$ and $y = p/q$. For the solution equations

$$\frac{dy}{dx} = \frac{r \cdot x \cdot y - s \cdot y}{p \cdot x - q \cdot x \cdot y} = \frac{rx - s}{x} \cdot \frac{y}{p - qy}$$

(We are assuming that $dx/dt \neq 0$; since $x > 0$, this amounts to the assumption that $y \neq p/q$.) Since $y > 0$, we can divide by y as part of a separation.

$$\frac{p - qy}{y} \cdot dy = \frac{rx - s}{x} \cdot dx$$

$$\left(\frac{p}{y} - q\right) \cdot dy = \left(r - \frac{s}{x}\right) \cdot dx$$

$$\int \left(\frac{p}{y} - q\right) \cdot dy = \int \left(r - \frac{s}{x}\right) \cdot dx$$

$$p \cdot \ln(y) - q \cdot y = r \cdot x - s \cdot \ln(x) + C$$

where C is a constant. It is quite interesting to graph some of these curves; in class we may have time to consider a couple of cases. ∎

---

[4]Simply: the wolves die off without food.

[5]Lotka and Volterra each developed this DE in a different context in the 1920's. See [**1**, p.529]

Kinetic energy and work.   It is frequently the case that the force exerted on a physical object depends on the object's position. Let's say we have an object on the y-axis and subject to an acceleration $f(y)$, so that

$$\frac{d^2y}{(dt)^2} = f(y)$$

Define $x = y'$ and we have the following two parametric equations.

$$\frac{dy}{dt} = x \quad \text{and} \quad \frac{dx}{dt} = f(y)$$

We can apply the trick we used above:

$$\frac{dy}{dx} = \frac{x}{f(y)}$$

When we separate x and y, and the resulting integrals give this:

$$\frac{x^2}{2} = \int f(y) \cdot dy$$

Recall that $x = y'$, and we have the mathematical derivation of a well-known principle of kinetics, for the quantity $x^2/2$ expresses the *kinetic energy* of the object, whereas the integral on the right is the *work done* in moving it. Thus, the equation we obtained is familiar to the physicist, for it says that the kinetic energy of the object is equal to the work done in moving it. ∎

Here is a problem where force depends on velocity rather than position.

Gravity and air resistance.   An object relatively close to the earth and falling straight down with velocity v (the positive direction is down) encounters a nearly constant gravitational acceleration g. We will make the convenient and not unreasonable assumption that the air resistance on the object is proportional to the velocity. That force, then, has the form $-\rho \cdot v$, where ρ is a positive constant. If the object has constant mass m, then the *acceleration* due to resistance is $-\rho \cdot v/m$. Write $k = \rho/m$ and we have acceleration $-k \cdot v$.

Taking both forces into account, the acceleration is this:

$$\frac{dv}{dt} = g - k \cdot v$$

Assuming $v_0 > 0$, let's solve for v and describe what happens to v as $t \to \infty$.

Notice that there is an equilibrium when $v = g/k$. When $v \neq g/k$, we can separate

$$dt = \frac{dv}{g - k \cdot v} \quad \text{so that} \quad \int_0^t dt = \int_{v_0} \frac{dv}{g - k \cdot v}$$

We integrate each side and continue.

$$t = -\frac{1}{k} \cdot \ln |g - k \cdot v| \Big|_{v_0}$$

$$t = -\frac{1}{k} \cdot \ln |g - k \cdot v| + \frac{1}{k} \cdot \ln |g - k \cdot v_0|$$

$$t = \frac{1}{k} \cdot \ln \left| \frac{g - k \cdot v_0}{g - k \cdot v} \right|$$

This can be solved[6] for v, but we can see what happens to v without solving! Letting $t \to \infty$, we see that the argument to the logarithm must go to infinity as well. The only way this can happen is to have the denominator go to 0. In other words $v \to g/k$. We see that the velocity approaches the equilibrium – that value is the *terminal velocity*. ■

Rocket thrust.   A rocket creates thrust by expelling burned fuel. The mass m of the rocket might be modeled as $m = m_0 - \alpha \cdot t$ where α is the *burn rate* in mass per unit time. It is a straightforward matter of physics that the burning fuel produces a force relative to the rocket of $\alpha \cdot (\beta - v)$, where v is the velocity. Let's assume also that the rocket travels through a medium offering resistance proportional to the velocity.[7] Thus, the resistance force is $-\rho \cdot v$ where ρ is a constant.

---

[6]You should think about resolving the absolute values: there are two cases: $v_0 < g/k$ and $v_0 > g/k$, but in both cases $(g - kv_0/(g - kv)$ is positive.

[7]This assumption is typical at low velocities. A homework problem explores resistance at higher velocities.

Since mass is not constant, we need to replace *force equals mass times acceleration* with the more general *force equals the time derivative of momentum.* We see that

$$\frac{dm}{dt} \cdot v + m \cdot \frac{dv}{dt} = \alpha \cdot (\beta - v) - \rho \cdot v$$

Using the formula for m and cancelling $\alpha \cdot v$ from both sides of the equation, we obtain the following.

$$(2.7) \qquad (m_0 - \alpha \cdot t) \cdot \frac{dv}{dt} = \alpha \cdot \beta - \rho \cdot v$$

Define $\gamma = \alpha \cdot \beta$, and we see that we have an equilibrium when $v = \gamma/\rho$. Otherwise, we can separate. (Note that $m \neq 0$, so we can divide by it.)

$$\frac{dv}{\gamma - \rho \cdot v} = \frac{dt}{m_0 - \alpha \cdot t}$$

$$\int_{v_0} \frac{dv}{\gamma - \rho \cdot v} = \int_0 \frac{dt}{m_0 - \alpha \cdot t}$$

$$-\frac{1}{\rho} \cdot \ln |\gamma - \rho \cdot v| \Big|_{v_0} = -\frac{1}{\alpha} \cdot \ln |m_0 - \alpha \cdot t| \Big|_0$$

$$\frac{1}{\rho} \cdot \ln \left| \frac{\gamma - \rho \cdot v_0}{\gamma - \rho \cdot v} \right| = \frac{1}{\alpha} \cdot \ln \left| \frac{m_0}{m_0 - \alpha \cdot t} \right|$$

We have to turn the engine off before t gets to m_0/α, or we won't have any rocket left! Imagining, however, that $m_0 - \alpha \cdot t$ is close to 0, its reciprocal is large and its logarithm is large as well. Looking at the left side, this necessitates that v be close to γ/ρ, the equilibrium. Thus, the equilibrium furnishes a theoretical upper limit on the velocity obtained by propulsion. ■

Soft Non-Linear Spring.   Here is a problem in which there are some interesting solution curves. Equations of this type were considered by Poincaré in the 1890's, see [**12**]. The equations describe what is called a *soft, non-linear spring.* We let $a > 0$, and we seek solution curves to the equation.

$$y'' = a \cdot y^3 - y$$

We write $x = y'$, and we have

$$\frac{dy}{dt} = x \quad \text{and} \quad \frac{dx}{dt} = \frac{d^2y}{(dt)^2} = a \cdot y^3 - y$$

Equilibria? We need $x = 0$ and $a \cdot y^3 - y = 0$, so that $y = 0, \pm\sqrt{1/a}$. To find solution curves, we consider

$$\frac{dy}{dx} = \frac{dy/dt}{dx/dt} = \frac{x}{a \cdot y^3 - y}$$

which separates:

$$\int (a \cdot y^3 - y) \cdot dy = \int x \cdot dx$$

$$\frac{a}{4} \cdot y^4 - \frac{y^2}{2} + C = \frac{x^2}{2}$$

You might try graphing some of these curves in the xy-plane. Suggestion: take $a = 1/4$, and the equilibria are $(0, 0)$ and $(0, 2)$ and $(0, -2)$. Consider various initial conditions: $x = 0, y = 1$ and $x = 2, y = 0$. (Each set of initial conditions will give you a specific value of C; graph away!) ∎

Logistic Population Growth. The population $P(t)$ of an animal in a definite area is limited by the size of the area. Suppose that the rate of change in P is jointly proportional to P (from reproductive growth) and to $E - P$ where E is a constant (from the limited area). If $E > P_0 > 0$, what happens as $t \to \infty$?

We have $P' = k \cdot P \cdot (E - P)$, where k is a positive constant. (This is called the *logistic equation*.) This has equilibria at $P = 0$ and $P = E$. Since $0 < P_0 < E$, we can separate. (The resulting integral in P needs the partial fractions formula from Chapter 1.) We'll put the dt on the left side.

$$k \cdot P \cdot (E - P) = \frac{dP}{dt}$$

$$\int_0 k \cdot dt = \int_{P_0} \frac{-dP}{P \cdot (P - E)}$$

$$k \cdot t = \frac{1}{E} \cdot \ln \left| \frac{P}{P - E} \right| \Big|_0 \qquad\qquad \text{formula p.2}$$

$$= \frac{1}{E} \cdot \ln \left| \frac{P}{P - E} \right| - \frac{1}{E} \cdot \ln \left| \frac{P_0}{P_0 - E} \right|$$

$$= \frac{1}{E} \cdot \ln \left(\frac{P}{E - P} \right) - \frac{1}{E} \cdot \ln \left(\frac{P_0}{E - P_0} \right)$$

(Notice that we resolved the absolute values on the logarithm. How?) This equation can be solved for P, but we can see what happens as $t \to \infty$ without solving. The log-term has to go to infinity, keeping $P < E$. The only way that can happen is to have $P \to E$, so that the population approaches equilibrium! This is an example of what is called a *stable equilibrium*, since if we start reasonably close to equilibrium, then we approach the equilibrium in the limit. ∎

3. First Order Linear Differential Equations

These equations look like this:

(2.8) $$\frac{dy}{dt} + q(t) \cdot y = g(t)$$

where $q(t), g(t)$ are continuous functions.[8] Although this pattern is simple, it may take some work with examples before you are confident you can recognize a first order linear problem reliably. We will move to examples very soon; first we explain a technique for solving these problems.

---

[8]It is possible to weaken this hypothesis considerably, but for the sake of simplicity in an introductory course, we will not pursue this weakening in any systematic way. However, a problem at the end of this chapter is suggestive.

The solution idea goes back to Euler; it appears in passing in a paper he published in 1728. (See [**6**, p.638] for an English translation of this paper.) There is a special function $\mu(t)$ such that if you multiply the equation by $\mu(t)$, the left side of the equation mimics the product rule. The special function is called the *integrating factor*, and it is defined like this.

$$\mu(t) = \exp\left(\int_0 q(t) \cdot dt\right)$$

To see the wisdom of this definition, you need to see how it can be used to turn a first order linear DE into a simpler integration problem. Its use in that regard hinges on the formula for its derivative. Remember that the derivative of a semi-definite integral is the integrand:

$$(2.9) \qquad \frac{d}{dt}\mu(t) = \exp\left(\int_0 q(t) \cdot dt\right) \cdot \frac{d}{dt}\int_0 q(t) \cdot dt = \mu(t) \cdot q(t)$$

In light of this, let's compute the derivative of $\mu \cdot y$:

$$\frac{d}{dt}(\mu(t) \cdot y) = \mu'(t) \cdot y + \mu(t) \cdot y'$$
$$= q(t) \cdot \mu(t) \cdot y + \mu(t) \cdot y'$$
$$= \mu(t) \cdot (q(t) \cdot y + y') \qquad\qquad \text{now use } (2.8)$$
$$= \mu(t) \cdot g(t)$$

Taking antiderivatives, we see that

$$(2.10) \qquad\qquad \mu(t) \cdot y = \int \mu(t) \cdot g(t) \cdot dt$$

The definition of $\mu(t)$ as an exponential function shows that it is never 0, and so we can divide by it to solve for y.

We can't resist speculating a little more about this solution idea. Euler probably started with the idea that a cleverly chosen μ might make the derivative of $\mu \cdot y$ come out nicely. If we do the calculation of $(\mu \cdot y)'$, we see that it simplifies if $\mu' = \mu \cdot q$. That formula is a separable DE, and we can solve for μ. We end up with μ being the integrating factor.

Back to practical matters. We want to be a little more precise about the solution we just sketched. Observe that $\mu(0) = 1$, and we integrate both sides of the equation $[\mu(t) \cdot y]' = \mu(t) \cdot g(t)$:

$$\int_0 \mu(t) \cdot g(t) \cdot dt = \int_0 \frac{d}{dt}\left(\mu(t) \cdot y\right) \cdot dt = \mu(t) \cdot y\Big|_0$$
$$= \mu(t) \cdot y - \mu(0) \cdot y_0 = \mu(t) \cdot y - y_0$$

We don't need to repeat the derivation every time; you can remember the following summary.

Solving a First Order Linear DE   For the equation

(2.11) $$\frac{dy}{dt} + q(t) \cdot y = g(t) \quad \text{with} \quad y_0$$

Define the *integrating factor*

$$\mu(t) = \exp\left(\int_0 q(t) \cdot dt\right)$$

and then

(2.12) $$\mu(t) \cdot y - y_0 = \int_0 \mu(t) \cdot g(t) \cdot dt$$

■

The derivation just summarized establishes both the existence of a solution to a first order linear DE and the uniqueness of that solution. Briefly, here are the details presented with attention to technicalities: suppose that $q(t), g(t)$ are continuous on an interval I on the real line, with 0 being in the interval. Then $\mu(t)$ exists for all t in I, since the continuous function $q(t)$ can be integrated there. If y is a solution to (2.11), then, as shown above, the equation (2.10) holds. Taking into account y_0, it is easy to see that (2.12) holds. Thus, there can be only one solution to (2.11). As to existence, we can *define* y so that (2.12) holds, since, as we have noted, $\mu(t)$ is never 0, and so it can be divided away from the left side to expose y. The derivation that produced (2.12) works backward to show that y satisfies (2.11). Thus, (2.11) has a unique solution defined in the interval I.

4. Examples and Applications

We begin with a straightforward use of (2.11).

Problem.   Solve the IVP.

$$y' + \frac{1}{t+2} \cdot y = t^2, \quad y_0 = 3$$

Solution.   Our integrating factor involves the integral

$$\int_0 \frac{dt}{t+2} = \ln|t+2| - \ln 2$$

We can eliminate the absolute values: since t starts at $t = 0$, where $t + 2 > 0$, we see that $t + 2$ stays positive.[9] And we can combine the logs.

$$\int_0 \frac{dt}{t+2} = \ln(t+2) - \ln(2) = \ln\left[\frac{t+2}{2}\right]$$

Now we are ready for the integrating factor formula.

$$\mu = \exp\left(\int_0 \frac{dt}{t+2}\right) = \exp\left(\ln\left[\frac{t+2}{2}\right]\right) = \frac{t+2}{2}$$

Then (2.12) is this:

$$\frac{t+2}{2} \cdot y - 3 = \int_0 \frac{t+2}{2} \cdot t^2 \cdot dt = \frac{t^4}{8} + \frac{t^3}{3}$$

So that

$$y = \frac{2}{t+2} \cdot \left(3 + \frac{t^4}{8} + \frac{t^3}{3}\right)$$

■

Our other examples will be application problems.

---

[9]The term $t+2$ occurs in the denominator, and so $t \neq -2$. Thus, if t starts with $t > -2$ it has to stay there.

Rocket Propulsion Revisited. We return to the rocket propulsion problem, including a constant downward gravitational acceleration and assuming that the rocket moves straight up.

We recall the equation (2.7), where the mass m is expressed as $m_0 - \alpha \cdot t$. Keeping this in mind, we include the gravitational force $-g \cdot m$.

$$m \cdot \frac{dv}{dt} = \alpha \cdot \beta - \rho \cdot v - g \cdot m$$

Dividing by the mass and rearranging, we obtain a first-order linear equation.

$$\frac{dv}{dt} + \frac{\rho}{m} \cdot v = \frac{\alpha\beta}{m} - g$$

We will leave it to you to check our work in computing the integrating factor. (You'll need to play with logarithms!)

$$\mu(t) = \exp\left(\int_0 \frac{\rho}{m} \cdot dt\right) = \left(\frac{m_0}{m}\right)^{\rho/\alpha}$$

Then

$$v \cdot \left(\frac{m_0}{m}\right)^{\rho/\alpha} - v_0 = \int_0 \left(\frac{m_0}{m}\right)^{\rho/\alpha} \cdot \left(\frac{\alpha\beta}{m} - g\right) \cdot dt$$

The integral on the right is messy but not hard: t is only involved in m, and we have powers of m when we multiply out. We've taken this as far as we want. You might observe that you can solve for v and then integrate again to get the position of the rocket! ■

Paying Back a Loan. Suppose we borrow P_0 dollars and pay it back at m dollars per month, incurring a finance charge (interest!) rate of r each month.[10] How long does it take us to pay back the loan?

If $P(t)$ is the amount left on the loan at time t, then the description of the loan tells us about $P(t + \Delta t)$, where Δt is one month.[11] It makes sense, then, to measure t in months. The finance charge makes the amount go up by $r \cdot P$,

---

[10]An example: If the annual interest rate is 6%, then $r = 0.06/12$.

[11]The amount $P(t)$ is the *principal*.

and then the payment makes the amount go down[12] by m. Here is the net change in P:

$$\Delta P = r \cdot P - m$$

Since $\Delta t = 1$ (month), we see that

$$\frac{\Delta P}{\Delta t} = r \cdot P - m$$

This equation can be solved exactly, but the solution is quite messy. It is easier to *approximate* $\Delta P / \Delta t \approx dP/dt$ and to think about the DE

$$\frac{dP}{dt} = r \cdot P - m \quad \text{which is} \quad \frac{dP}{dt} - r \cdot P = -m$$

The second equation is a first order linear equation.[13]

Integrating factor: $\mu = \exp(\int_0 (-r) \cdot dt) = \exp(-rt)$, and we have

$$\exp(-rt) \cdot P - P_0 = \int_0 \exp(-rt) \cdot (-m) \cdot dt = \frac{m}{r}\left(e^{-rt} - 1\right)$$

and so, simplifying some, we obtain

(2.13) $$P = \frac{m}{r} + e^{rt} \cdot \left(P_0 - \frac{m}{r}\right)$$

We want P to decrease! Thus, we better have $P_0 < m/r$, so that the e^{rt} term decreases. This in itself says something interesting: given monthly payment m and monthly interest rate r, the ratio m/r is the theoretical maximum you can borrow. We wanted to know when $P = 0$. Solving (2.13) in that case, we obtain

$$t = \frac{1}{r} \cdot \ln\left(\frac{m}{m - r \cdot P_0}\right)$$

---

[12]We're assuming that the finance charge is applied before the payment is applied. You can apply the payment first; the difference between the two ways is usually not significant.

[13]The equation is also separable! You might compare the solutions obtained from the two techniques.

Another interesting problem: find monthly payment m given P_0 and r and the desired number of months t for the loan. Some algebra shows that

$$m = \frac{e^{rt} \cdot r \cdot P_0}{e^{rt} - 1}$$

∎

Response to Poly-Exponential Stimulus.   There are a great many DE's that describe the response of a physical system to external stimuli. That last sentence is so general as to be almost useless – we will see specific examples in a variety of contexts as we travel through the course. For now we consider a system represented by the function $y(t)$ and a constant α. When an external stimulus $f(t)$ occurs, the following first order linear DE describes the response:

$$\frac{dy}{dt} - \alpha \cdot y = f(t)$$

A specific example: the gravity-air resistance equation was of this form. But right now, we are more interested in what y looks like when $f(t)$ is poly-exponential. This information will be needed in Chapter 7. Here is what we need; it is very general in that the constant α can be complex.

THEOREM 2.1. *Let $f(t)$ be a poly-exponential function, and let α be a complex number. Then the following IVP has a unique solution, and that solution is poly-exponential:*

$$\frac{dy}{dt} + \alpha \cdot y = f(t) \quad \text{given} \quad y_0$$

Furthermore, if $f(t)$ is a polynomial, then there is a value of y_0 such that the solution is a polynomial.

PROOF. Since the equation is first-order linear, it has a unique solution, as discussed above. The point is that the solution is poly-exponential. To see this, notice first that the integrating factor is $\mu(t) = \exp(\alpha \cdot t)$. Then the formula for the solution to the IVP can be deduced from (2.12).

$$y = \exp(-\alpha \cdot t) \left(\int_0 \exp(\alpha \cdot t) \cdot f(t) \cdot dt + y_0 \right)$$

If the semi-definite integral is poly-exponential, then y will be poly-exponential, as needed. To see that the semi-definite integral is poly-exponential, write $f(t)$ in its poly-exponential form:

$$f(t) = \sum_{k=1}^{n} P_k(t) \cdot \exp(\beta_k \cdot t)$$

where $P_1(t), \ldots, P_n(t)$ are polynomials and $\beta_1, \ldots, \beta_n$ are complex numbers. Here is the semi-definite integral:

$$\int_0 \exp(\alpha \cdot t) \cdot f(t) \cdot dt = \int_0 \exp(\alpha \cdot t) \cdot \sum_{k=1}^{n} P_k(t) \cdot \exp(\beta_k \cdot t) \cdot dt$$

$$= \int_0 \sum_{k=1}^{n} P_k(t) \cdot \exp((\beta_k + \alpha) \cdot t) \cdot dt$$

As noted in Proposition 1.3 in Chapter 1, the integral of each term $P_k(t) \cdot \exp((\beta_k + \alpha) \cdot t)$ is poly-exponential. If $f(t)$ is a polynomial, then we may as well assume that $f(t) = P_1(t)$ with $\beta_1 = 0$. We see that the $\exp(\alpha \cdot t)$ terms cancel in the formula for y, except at the constant term. We can adjust y_0 to make this term 0. $\qquad \square$

5. Problems

2.1. Solve the following IVP's. In each case be deliberate about classifying the problem as separable or first order linear. If logarithms occur, write their arguments appropriately without absolute values.

a) $x' + \dfrac{7}{t-5} \cdot x = 2$, $x_0 = 0$ **b)** $y' + \cot(t) \cdot y = 2 \cdot \cos(t)$, $y(\pi/2) = 0$

c) $A' + t^3 \cdot A^2 = 9 \cdot t^3$, $A_0 = 1$ **d)** $t \cdot u' + 3 \cdot u = t$, $u(1) = 0$

e) $y' = \cos(y) \cdot \tan^5(t)$, $y_0 = \pi/2$ **f)** $z' = z^3$, $z_0 = 1$

g) $y' + \dfrac{2t}{t^2+1} \cdot y = t^2$, $y_0 = 1$ **h)** $q' - 7 \cdot e^t \cdot q = 0$, $q_0 = -1$

i) $w' - \exp(2 \cdot w) \cdot t = 0$, $w_0 = 0$ **j)** $y' + 4 \cdot y = 2 \cdot t$, $y_0 = 0$

k) $x' + 3 \cdot t^2 \cdot x = t^6$, $x_0 = 1$ **λ)** $z' = \dfrac{e^z \cdot \sin(3t)}{z}$, $z_0 = 1$

2.2. Solve for x in this IVP, and determine the *domain* (the set of possible values of t).

$$\frac{dx}{dt} = t \cdot \sqrt{9 - x^2} \quad \text{and} \quad x_0 = 0$$

2.3. Solve this first-order linear IVP.

$$x'' - 3 \cdot x' = 2 \cdot e^{3t} \quad \text{and} \quad x_0 = 1,\ x_0' = 1$$

2.4. Consider the *Bernoulli Equation*: $x' + p(t) \cdot x + q(t) \cdot x^n = 0$. Make the substitution $w = x^{1-n}$ and see what happens.

2.5. We will show that there is a continuous solution y to this *discontinuous* IVP:

$$y' - 3 \cdot y = \begin{cases} 2 & \text{when } 0 \le t \le 1 \\ 5 & \text{when } 1 < t \end{cases} \quad \text{and} \quad y_0 = 0$$

(a) Solve for y when $0 \le t \le 1$. Take note of $y(1)$.

(b) Solve for y when $t > 1$, using $y(1)$ as initial value.

(c) Note that y is continuous for all $t \ge 0$. Describe the graph of y at $t = 1$.

2.6. Modify the gravity and air resistance model on p.18: replace $k \cdot v$ by $k \cdot v^2$. (This follows the *drag equation* for rockets at high velocity.) Assuming that we keep $v > 0$ (that we are moving down), find the equilibrium. If v_0 starts below equilibrium, what happens as $t \to \infty$?

2.7. Suppose that a population P has $P_0 > 0$ and grows at a rate proportional to the m-th power of the population. (Exponential growth occurs when $m = 1$.)

a) Suppose that $0 < m < 1$. Show that $P \to \infty$ as $t \to \infty$.

b) Suppose that $1 < m$. Show that $P \to \infty$ as $t \to A$ for some positive number A. (We say that P *blows up in finite time*.) Make sure $A > 0$.

2.8. Placing a cake in the oven or a soda can in a refrigerator are special cases of the following: an object with no heat source of its own is placed in an environment of constant temperature E. What is called *Newton's law of cooling*[14] asserts that the object's temperature will change at a rate proportional to the difference between E and the object's temperature. Let y stand for the temperature of the object as a function of time t, and assume that $0 \le y_0 < E$. Show that $y \to E$ as $t \to \infty$. In the specific case that $E = 350°$F and $y_0 = 70°$F, suppose that $y = 200°$F after half an hour. What is y after an hour and a quarter?

2.9. In the Lotka-Volterra equation, let $p = q = r = s = 1$, so that the equilibria at $(0,0)$ and $(1,1)$. (Not *one rabbit, one wolf* but one unit of population for each!) Graph the solution curve having $x_0 = 1/2$ and $y_0 = 1/2$. (Hint: find C based on the initial conditions. The graph is hard to get: factor $x' = x(1 - y)$ and think about whether x increases or decreases; do the same for y.)

---

[14]If the temperature of the object is less than E, it will heat up, and *cooling* becomes *heating*. Mathematicians amuse themselves by thinking of heating as *negative cooling*; physically the two processes are obviously very different!

2.10. In the logistic equation, it is of mathematical interest to consider an initial value $P_0 < 0$. Show that P *blows down* (goes to $-\infty$) in finite time. (Note: make sure you show that the time is positive when $P = \infty$.)

2.11. *(A simple economic model.)* Suppose that the price p of a good changes at a rate proportional to the difference between the demand and supply of that good. Assume that demand is a decreasing, linear function of p, and that supply is an increasing linear function of p. Get a first order linear DE out of this.

2.12. Modify the loan-payment equation for this problem: we invest m dollars per month in an account that grows at a monthly interest rate r. Solve the equation for P, and determine the *doubling time:* the value of t when $P = 2 \cdot P_0$.

2.13. Solve the rocket propulsion equation in the (easier!) case that $\rho = 0$. Show that the velocity of the rocket can be made greater than the velocity of the exhaust gasses. (This fact is important in designing a rocket to escape the gravitational pull of a planet.)

2.14. An object is placed on the x-axis at the point $x_0 > 0$ with zero initial velocity. An inverse square gravitational force acts on the object, pulling it toward the "sun" at the origin. Write $y = x'$ and get a DE by writing x' and y' in terms of x and y. Now use the equation for dy/dx to solve for y as a function of x. Remembering that $y = x'$, separate x and t, write t as an integral in x. (The integral can be done, and if you are ambitious you might try it, but you can leave the integral alone if you wish.)

2.15. A tank holds 10 gallons of water and the water is capable of holding a great deal of salt. Initially there is 1 gallon of water and no salt in the tank. Then two spigots are turned on, so that water enters the tank at 3 gallons per minute and salt enters at 2 ounces per minute. The tank is constantly stirred, as well, so that the salt mixes thoroughly with the water without changing the volume of the water. When the water and salt begin to enter, a tap opens at the bottom of the tank and lets out 1 gallon of salt-water per minute. Write down a DE that describes the amount of salt in the tank at time t. (Hint: first solve for the amount w of water as a function of time; let the units on salt and water guide you.) Solve the DE you wrote down; how much salt is in the tank when the tank is full?

2.16. It is a very important physical principle that systems tend to respond to a periodic stimulus in the same frequency as the stimulus. If f is constant, then the function $\cos(2 \cdot \pi \cdot f \cdot t)$ has frequency f. Show that the solution to

$$x' + k \cdot x = \cos(2 \cdot \pi \cdot f \cdot t), \quad x_0 = 0$$

has frequency f, as well. (You can assume that k is a positive constant.)

CHAPTER 3

Matrix Algebra

A *matrix* is a table of numbers.[1] Such a table organizes its entries into horizontal rows and vertical columns. The entries are often enclosed in parentheses:

$$\begin{pmatrix} 1 & 2 & 3 \\ 4 & 5 & 6 \end{pmatrix}$$

This matrix has 2 rows and 3 columns, and so we say it has *size* 2×3, and we say it is a 2×3 matrix. The rows and columns are numbered in an obvious way: rows from top to bottom and columns from left to right. Thus, the second row of the given matrix is $\begin{pmatrix} 4 & 5 & 6 \end{pmatrix}$ and the first column is $\begin{pmatrix} 1 \\ 4 \end{pmatrix}$. The entry in row i and column j is called the *i, j-entry*.

More generally (more abstractly), if m and n are positive integers, an $m \times n$ matrix M has i, j-entry $M[i, j]$ for all $1 \leq i \leq m$ and $1 \leq j \leq n$. You will often see subscripts employed to designate entries: $M_{i,j}$ instead of $M[i, j]$; we prefer the more readable bracket notation $M[i, j]$.

A matrix with a single row or a single column is often called a *vector*. For instance, if M is $1 \times n$, then since M has n real number entries, we can think of M as an element of $\mathbb{R}^n$. In vector notation, the coordinate M_i (which we prefer writing $M[i]$) is the matrix entry $M[1, i]$. Because there is only one row, the 1 in $M[1, i]$ is not really necessary. Similarly, if M is $n \times 1$, then $M[i, 1]$ can be thought of as $M[i]$, the i-th coordinate of M as an element of $\mathbb{R}^n$.

---

[1]For "numbers" we can usually stick to real numbers (we can often stick to integers!), although, as we will see, complex numbers come up quite naturally in matrix applications.

This course introduces the many applications of matrices. Quite often, an innocent looking table of numbers takes on *algebraic significance* in a particular application – we will see examples very shortly. For now, we define the algebraic operations that will occur in applied problems.

First, we need to be able to tell matrices apart. Two matrices are *equal* if they have the same size and if corresponding entries are equal (as numbers). This may seem trivial or obvious, but it carries some important subtleties. You might get the impression, at first, that this definition of equality is primarily a matter of bookkeeping. Hold that thought.

1. Matrix Arithmetic

Matrices can be added in a manner similar to the way vectors are added. If A and B are $m \times n$ matrices (we are saying that A and B have the same size), then the matrix $A + B$ is defined to be $m \times n$ and

$$(A + B)[i,j] = A[i,j] + B[i,j] \quad \text{for all} \quad 1 \le i \le m, \ 1 \le j \le n$$

In other words, you add matrices the same way you add vectors: entry by entry. For instance

$$\begin{pmatrix} 1 & 2 & 3 \\ 4 & 5 & 6 \end{pmatrix} + \begin{pmatrix} -2 & 3 & 7 \\ 5 & -6 & -6 \end{pmatrix} = \begin{pmatrix} -1 & 5 & 10 \\ 9 & -1 & 0 \end{pmatrix}$$

If A and B have different sizes, then their sum is not defined (and not worth thinking about).

It is obvious but worth verifying that addition of matrices is *commutative*: $A + B = B + A$ for all matrices A and B of the same size. Let's be formal about this; we will use it as a first occasion to remind ourselves what it means for matrices to be equal. To show that $A + B = B + A$, we need to see that $A + B$ and $B + A$ have the same size, and then that they have the same entries: that $(A + B)[i,j] = (B + A)[i,j]$ for all relevant i and j.

The definition of matrix addition makes $A + B$ and $B + A$ the same size as A, B, and so $A + B$ and $B + A$ are the same size as each other. As for entries, if $1 \leq i \leq m$ and $1 \leq j \leq n$, then

$$
\begin{aligned}
(A + B)[i, j] &= A[i, j] + B[i, j] &&\text{definition of } A + B \\
&= B[i, j] + A[i, j] &&\text{real addition is commutative} \\
&= (B + A)[i, j] &&\text{definition of } B + A
\end{aligned}
$$

It is important that you understand the different uses of the plus symbol $+$ in the foregoing. When we write $A + B$, the plus sign stands for matrix addition, and we are thinking of $A + B$ as one matrix formed from A, B. The expression $A[i, j] + B[i, j]$ uses the plus sign in its ordinary meaning as real number addition. It is tempting to confuse the real number entry $A[i, j]$ with the matrix A. It will help your understanding to distinguish them attentively.

Similarly, matrix addition is *associative*: if A, B, C are matrices all of the same size, then

$$(A + B) + C = A + (B + C)$$

We can give a proof this property that is similar to the proof that addition is commutative.

It is easy to take such things as the commutative and associative properties for granted; that is precisely the point of having them. These identities allow us to perform complicated matrix additions using familiar properties of numbers.

A "zero" is an additive identity element: an element which has no effect when it is added. The real number 0 is the additive identity element in real addition; the zero vector is the additive identity element in vector addition. The *zero matrix* is the matrix which has all its entries zero. Since there are various possible sizes for matrices, there is a zero matrix of each size. We will write $\mathbb{O}_{m \times n}$ for the $m \times n$ zero matrix. Because matrix addition is done entry by entry, it is obvious that the zero matrix is an additive identity: if A is an

$m \times n$ matrix, then

$$A + \mathbb{O}_{m \times n} = A = \mathbb{O}_{m \times n} + A$$

The next thing we want to do is to multiply a matrix by a number. Just as in scalar multiplication of a vector by a number, each matrix entry is multiplied by the number (scalar) in question. Thus, for an $m \times n$ matrix A and a number α, we define the $m \times n$ matrix $\alpha \cdot A$ where

$$(\alpha \cdot A)[i, j] = \alpha \cdot A[i, j] \quad \text{for all} \quad 1 \le i \le m, \ 1 \le j \le n$$

We pause to call attention to the two uses of the multiplication dot: in $(\alpha \cdot A)$ it stands for scalar multiplication of the matrix A; in $\alpha \cdot A[i, j]$ it refers to real number multiplication.

As with matrix addition, the properties of scalar multiplication are obvious and easy to verify. We want to list them to be complete and to give you a chance to pick a couple of them to prove, making sure you really understand what is meant in the definitions.

a) If A, B are $m \times n$ matrices and α is a number, then $\alpha \cdot (A + B) = (\alpha \cdot A) + (\alpha \cdot B)$.

b) If A is $m \times n$ and α, β are numbers, then $(\alpha + \beta) \cdot A = (\alpha \cdot A) + (\beta \cdot A)$.

c) If A is $m \times n$, then $0 \cdot A = \mathbb{O}_{m \times n}$ and $1 \cdot A = A$.

Playing with these identities yields the "negative" of a matrix. Indeed, compute:

$$
\begin{aligned}
\mathbb{O}_{m \times n} &= 0 \cdot A & \\
&= (1 + (-1)) \cdot A & \text{by (c)} \\
&= 1 \cdot A + (-1) \cdot A & \text{using (b)} \\
&= A + (-1)A & \text{by (c)}
\end{aligned}
$$

The point is that A and $(-1)A$ add up to the zero matrix. As you might expect, we write $-A$ for $(-1)A$. Notice that the entries of $-A$ are the negatives of the entries of A.

We need one more operation on matrices, a kind of multiplication. The formula we will give is not as easy to motivate from the get-go – later you will see that a matrix can be viewed as a function, and then matrix multiplication will be seen to correspond to function composition. For our immediate purposes, it will be better just to introduce multiplication and begin getting used to how it is computed.

To give the definition we need to recall the definition of the *dot product of vectors*. For matrix multiplication, one of the vectors will be a row and the other a column. Thus, we will speak of the *dot product of a row and a column*. Here is an example which models the general case and in which the symbol $\circ$ is used for the dot product.

$$\begin{pmatrix} 1 & 0 & -2 & 3 \end{pmatrix} \circ \begin{pmatrix} 3 \\ 4 \\ 1 \\ -5 \end{pmatrix} = (1 \cdot 3) + (0 \cdot 4) + (-2 \cdot 1) + (3 \cdot (-5))$$

$$= 3 + 0 - 2 - 15 = -14$$

In general,

$$\begin{pmatrix} a_1 & a_2 & \cdots & a_n \end{pmatrix} \circ \begin{pmatrix} b_1 \\ b_2 \\ \vdots \\ b_n \end{pmatrix} = a_1 b_1 + a_2 b_2 + \cdots a_n b_n = \sum_{i=1}^{n} a_i b_i$$

Now we are ready to define matrix multiplication. There are several nuances in this definition, and so you need to make sure you understand (and can reproduce!) the details. Given a matrix A that is $m \times n$ and a matrix B that is $n \times r$, we can define $A \cdot B$ and it will be $m \times r$. In other words, for the matrix product $A \cdot B$ to be defined, the matrix A has to have as many columns as B has rows. The best way to check this is to put the sizes of A and B side by side (keep A's size on the left),

$$m \times n \quad n \times r$$

and to check that the numbers in the middle (the n's) are the same. The size of the product $A \cdot B$ is obtained from the outside numbers left to right $m \times r$.

Just from the condition on the sizes, we see that matrix multiplication is not commutative. If A is 2×3 and B is 3×4, then $A \cdot B$ is defined (it is 2×4), but $B \cdot A$ is not defined. Another example: if A is 3×5 and B is 5×3, then both $A \cdot B$ and $B \cdot A$ are defined, but they cannot be equal because they do not have the same size. (What is the size in each case?)

Back to the definition. Let A be $m \times n$ and B $n \times r$. Then $A \cdot B$ is defined and it is $m \times r$. We have

$$(A \cdot B)[i, j] = (\text{row } i \text{ of } A) \circ (\text{column } j \text{ of } B)$$

$$\text{for all } 1 \leq i \leq m, \ 1 \leq j \leq r$$

It is a matter of bookkeeping to notice that row i of A has n entries (one entry for each column of A) and column j of B has n entries (one entry for each row of B). Thus, the indicated dot product makes sense. This defines matrix multiplication.

Consider

$$\begin{pmatrix} 1 & 2 & 3 & 0 \\ 4 & 5 & 6 & 0 \\ 7 & 8 & 9 & 0 \end{pmatrix} \cdot \begin{pmatrix} 1 & -1 & 2 \\ 3 & 0 & 4 \end{pmatrix}$$

Oops, the product does not make sense. (Why not?) Reverse the order.

$$\begin{pmatrix} 1 & -1 & 2 \\ 3 & 0 & 4 \end{pmatrix} \cdot \begin{pmatrix} 1 & 2 & 3 & 0 \\ 4 & 5 & 6 & 0 \\ 7 & 8 & 9 & 0 \end{pmatrix}$$

Now the product is defined, and it will be 2×4, and so we can set up a template for the answer.

$$\begin{pmatrix} 1 & -1 & 2 \\ 3 & 0 & 4 \end{pmatrix} \cdot \begin{pmatrix} 1 & 2 & 3 & 0 \\ 4 & 5 & 6 & 0 \\ 7 & 8 & 9 & 0 \end{pmatrix} = \begin{pmatrix} *_{1,1} & *_{1,2} & *_{1,3} & *_{1,4} \\ *_{2,1} & *_{2,2} & *_{2,3} & *_{2,4} \end{pmatrix}$$

We have indicated the row and column numbers of each entry in the product. Let's compute a few random entries. The $2, 3$-entry on the right is the dot product of row 2 of the first factor on the left with column 3 of the second factor.

$$*_{2,3} = \begin{pmatrix} 3 & 0 & 4 \end{pmatrix} \circ \begin{pmatrix} 3 \\ 6 \\ 9 \end{pmatrix} = 9 + 36 = 45$$

The $1, 4$-entry on the right is the dot product of row 1 of the first factor on the left with column 4 of the second factor.

$$*_{1,4} = \begin{pmatrix} 1 & -1 & 2 \end{pmatrix} \circ \begin{pmatrix} 0 \\ 0 \\ 0 \end{pmatrix} = 0$$

And so on. Here is the complete answer.

$$\begin{pmatrix} 1 & -1 & 2 \\ 3 & 0 & 4 \end{pmatrix} \cdot \begin{pmatrix} 1 & 2 & 3 & 0 \\ 4 & 5 & 6 & 0 \\ 7 & 8 & 9 & 0 \end{pmatrix} = \begin{pmatrix} 11 & 13 & 15 & 0 \\ 31 & 38 & 45 & 0 \end{pmatrix}$$

Going back to the definition of matrix multiplication, we want to write out what an individual entry of the product looks like – in addition notation and in summation notation. Let A be $m \times n$ and let B be $n \times r$, and then

$$(A \cdot B)[i, j] = A[i, 1]B[1, j] + A[i, 2]B[2, j] + \cdots + A[i, n]B[n, j]$$

$$= \sum_{k=1}^{n} A[i, k]B[k, j]$$

The terms $A[i, k]$ for $1 \leq k \leq n$ are the entries in row i of A, and the terms $B[k, j]$ for $1 \leq k \leq n$ form the j-th column of B. These formulas are extremely useful in dealing with matrix multiplication generally, as we will see shortly.

The definition of the matrix product $A \cdot B$ involves the rows of A and the columns of B. Here is a way to extract that idea. Let A be $m \times n$ and let B be $n \times r$, and let $1 \leq i \leq m$. We can view row i of A as a $1 \times n$ matrix A_i. We claim that $A_i \cdot B = (A \cdot B)_i$ – the product of the i-th row of A and B is the i-th row of $A \cdot B$. Notice that A_i is $1 \times n$, and so $A_i \cdot B$ is defined and $1 \times r$.

Since $A \cdot B$ is $m \times r$, the matrix $(A \cdot B)_i$ is $1 \times r$, the same size as $A_i \cdot B$. The $1, j$ entry of $A_i \cdot B$ is

$$A_i \circ (\text{ column } j \text{ of } B)$$

Since A_i is the i-th row of A, we see that our dot product is the i, j entry of $A \cdot B$, and that's the $1, j$ entry of $(A \cdot B)_i$.

A similar fact holds for columns. Keeping the same A, B as in the previous paragraph, this time, use subscripts for columns. Thus, B_k is the k-th column of B, considered as an $n \times 1$ matrix. We claim that $A \cdot B_k = (A \cdot B)_k$. We will leave the argument to you or to class. Notice that $A \cdot B_k$ and $(A \cdot B)_k$ are $m \times 1$.

We have shown, on the grounds of size alone, that matrix multiplication is not commutative. Consider the following.

$$\begin{pmatrix} 0 & 1 \\ 0 & 0 \end{pmatrix} \cdot \begin{pmatrix} 0 & 0 \\ 1 & 0 \end{pmatrix} = \begin{pmatrix} 1 & 0 \\ 0 & 0 \end{pmatrix} \neq \begin{pmatrix} 0 & 0 \\ 0 & 1 \end{pmatrix} = \begin{pmatrix} 0 & 0 \\ 1 & 0 \end{pmatrix} \cdot \begin{pmatrix} 0 & 1 \\ 0 & 0 \end{pmatrix}$$

This shows, even when $A \cdot B$ and $B \cdot A$ are defined and the same size, they are not necessarily equal.

Even though matrix multiplication is not commutative, it is associative. This fact is not an abstract amusement; we will see that it has applied consequences, and so it merits a careful statement and proof. Here goes. If A is $m \times n$ and B is $n \times r$ and C is $r \times s$, then

$$(A \cdot B) \cdot C = A \cdot (B \cdot C)$$

To prove this, we need to show that the matrices on each side of the equation are defined, are the same size, and are equal entry by entry.

As for sizes,

$$\begin{array}{ccc} A & B & C \\ m \times n & n \times r & r \times s \end{array}$$

On the left, A is $m \times n$ and B is $n \times r$, and so $A \cdot B$ is defined to be $m \times r$. Then $A \cdot B$ is $m \times r$ and C is $r \times s$, we see that $(A \cdot B) \cdot C$ is defined and it is $m \times s$. You should show that $A \cdot (B \cdot C)$ is defined as well, and that it is also $m \times s$. The proof that the entries of $(A \cdot B) \cdot C$ are those of $A \cdot (B \cdot C)$ is included at the end of this section.

Number addition and multiplication are connected by the distributive law. For matrices, if A and B are $m \times n$, and if C is $n \times r$, then

$$(3.1) \qquad (A + B) \cdot C = (A \cdot C) + (B \cdot C)$$

There is a similar identity

$$(3.2) \qquad D \cdot (E + F) = (D \cdot E) + (D \cdot F)$$

Suppose that D is $m \times n$, and you should be able to describe the sizes of E and F which make the equation defined. We will establish the entry-by-entry equality in (3.1) and (3.2) partly in class and partly in homework.

Scalars move around freely in matrix multiplication. Let A be $m \times n$, let B be $n \times r$, and let α be a number. Then

$$(\alpha \cdot A) \cdot B = A \cdot (\alpha \cdot B) = \alpha \cdot (A \cdot B)$$

As with the other basic facts, we'll leave this to class or to your work.

The zero matrices are additive identity elements, the *identity matrices* are multiplicative identity elements. Here they are:

$$I_2 = \begin{pmatrix} 1 & 0 \\ 0 & 1 \end{pmatrix}, \quad I_3 = \begin{pmatrix} 1 & 0 & 0 \\ 0 & 1 & 0 \\ 0 & 0 & 1 \end{pmatrix}, \quad I_4 = \begin{pmatrix} 1 & 0 & 0 & 0 \\ 0 & 1 & 0 & 0 \\ 0 & 0 & 1 & 0 \\ 0 & 0 & 0 & 1 \end{pmatrix}, \quad \text{etc.}$$

For a positive integer n, the matrix I_n, the $n \times n$ *identity matrix* has 1's along its *diagonal* and 0's everywhere else. To be more precise:

$$I_n[i, j] = 0 \quad \text{when} \quad i \neq j$$

and $\qquad I_n[i, i] = 1 \quad \text{where} \quad 1 \leq i \leq n$

The I_n are multiplicative identities; they have no effect when they are used to multiply. Let A be $m \times n$, and we will see that multiplication by an identity matrix results in A. Notice that the size of the identity matrix we use depends on whether we multiply on the left of A or on the right of A, because of the size requirement of matrix multiplication. Here are the two equations:

$$I_m \cdot A = A \quad \text{and} \quad A \cdot I_n = A$$

We will verify the second of these equations in two steps, the first of which is useful on its own.

PROPOSITION 3.1. *Let A be an $m \times n$ matrix, and choose j with $1 \le j \le n$. Let E_j be the j-th column of I_n. Then $A \cdot E_j$ is the j-th column of A.*

PROOF. The definition of I_n shows that E_j is $n \times 1$, and so $A \cdot E_j$ is defined, and it has size $m \times 1$. That's the size of a column of A.

Again, thinking of the form of E_j, we see that $E_j[k, 1] = 0$ when $k \ne j$, and $E_j[j, 1] = 1$. Here is the computation of an entry of $A \cdot E_j$.

$$(A \cdot E_j)[i, 1] = (\text{row } i \text{ of } A \circ E_j) = \sum_{k=1}^{n} A[i, k] \cdot E_j[k, 1]$$

In the sum, k moves from 1 to n while i stays fixed. We have $E_j[k, 1] = 0$ except when $k = j$. Thus,

$$(A \cdot E_j)[i, 1] = A[i, j]$$

We see that $A \cdot E_j$ is the j-th column of A. □

A corollary: $A \cdot I_n = A$. Indeed, we have already called attention to the size of $A \cdot I_n$; make sure you see that it is the same as the size of A. We showed above that the j-th column of $A \cdot I_n$ is $A \cdot E_j$. Proposition 3.1 says that $A \cdot E_j$ is the j-th column of A. We conclude that $A \cdot I_n$ and A have the same columns, and so they are equal.

There is one more topic to introduce. Given the $m \times n$ matrix A, its *transpose*, denoted A^T, is defined by changing the rows of A into columns: $A^T[i, j] = A[j, i]$ for all i, j. Thus, A^T is $n \times m$. We also have the following properties, which may be left as exercises.

(a) $(A^T)^T = A$;

(b) $(A + B)^T = A^T + B^T$ for all $m \times n$ matrices A, B.

(c) If A is $m \times n$ and B is $n \times r$, then $(A \cdot B)^T = B^T \cdot A^T$.

We have mentioned that $n \times 1$ matrices can be thought of as vectors – as elements of $\mathbb{R}^n$. If u, v are $n \times 1$, observe that $u^T \cdot v$ is the dot product of u and v. Recall that if you take the dot product of a vector with itself, you get the sum of squares of its entries:

$$u \circ u = u^T \cdot u = \sum_{j=1}^{n} u[j]^2$$

This sum is often denoted $|u|^2$, the square of the *norm* or *length* of u. Because the entries of u are real numbers, we have $|u|^2 \geq 0$, and if $|u| = 0$, then all the entries of u are 0. This property is called *positive definiteness* and it figures into many applications.

We close this section with a summary of the algebraic properties of matrix arithmetic. Throughout the following list, let A be an $m \times n$ matrix, and let B, C be matrices of the appropriate size, in each case, so that the operations involved make sense. Let α, β be numbers.

(1) $A + B = B + A$

(2) $(A + B) + C = A + (B + C)$

(3) $A + \mathbb{O}_{m \times n} = A$

(4) $\alpha \cdot (A + B) = (\alpha \cdot A) + (\alpha \cdot B)$

(5) $(\alpha + \beta) \cdot A = (\alpha \cdot A) + (\beta \cdot A)$

(6) $\alpha \cdot (\beta \cdot A) = (\alpha \cdot \beta) \cdot A$

(7) $(A \cdot B) \cdot C = A \cdot (B \cdot C)$

(8) $I_m \cdot A = A$ and $A \cdot I_n = A$

(9) $A \cdot (B + C) = (A \cdot B) + (A \cdot C)$

(10) $(A + B) \cdot C = (A \cdot C) + (B \cdot C)$

(11) $(\alpha \cdot A) \cdot B = A \cdot (\alpha \cdot B)$

We have mentioned the importance of the associative law of multiplication. For completeness, we here provide a proof. Let A be $m \times n$, let B be $n \times r$, let C be $r \times s$, and we claim that property (7) holds. In class, we showed that each side of the equation has size $m \times s$, so we need only show that corresponding entries are equal. Here goes: let $1 \le i \le m$ and $1 \le j \le s$. Then since $(A \cdot B)$ is $m \times r$ and C is $r \times s$, we have

$$\Big((A \cdot B) \cdot C\Big)[i,j] = \sum_{k=1}^{r} (A \cdot B)[i,k] \cdot C[k,j]$$

$$= \sum_{k=1}^{r} \left[\sum_{p=1}^{n} A[i,p] \cdot B[p,k] \right] \cdot C[k,j] \qquad \text{definition } (AB)[i,k]$$

$$= \sum_{k=1}^{r} \sum_{p=1}^{n} A[i,p] \cdot B[p,k] \cdot C[k,j] \qquad \text{associative law on } \mathbb{R}$$

$$= \sum_{p=1}^{n} \sum_{k=1}^{r} A[i,p] \cdot B[p,k] \cdot C[k,j] \qquad \text{commutative law on } \mathbb{R}$$

$$= \sum_{p=1}^{n} A[i,p] \cdot \sum_{k=1}^{r} B[p,k] \cdot C[k,j] \qquad \text{factor out } A[i,p]$$

$$= \sum_{p=1}^{n} A[i,p] \cdot (B \cdot C)[p,j] \qquad \text{definition of } (BC)[p,j]$$

$$= \Big(A \cdot (B \cdot C)\Big)[i,j]$$

and we're done.

2. An Application: Rotation in the Plane

We imagine rotating points in the plane about the origin through some given angle θ. If we start at (x, y), and if this point is rotated, then the angle addition formulas[2] can be used to show that we end up at the point

$$(\, x\cos(\theta) - y\sin(\theta) \, , \, x\sin(\theta) + y\cos(\theta) \,)$$

Do you see a matrix multiplication? Observe that

(3.3) $$\begin{pmatrix} \cos(\theta) & -\sin(\theta) \\ \sin(\theta) & \cos(\theta) \end{pmatrix} \begin{pmatrix} x \\ y \end{pmatrix} = \begin{pmatrix} x\cos(\theta) - y\sin(\theta) \\ x\sin(\theta) + y\cos(\theta) \end{pmatrix}$$

We have written the coordinates of points in the plane vertically rather than horizontally; this is a convention we will observe often in this course.

The matrix on the left of (3.3) is the *rotation matrix* for the angle θ, we denote it by $R(\theta)$. Equation (3.3) can be stated like this: to rotate the point (x, y), multiply it on the left by the rotation matrix. In other words, the rotation matrix is the *function* "rotate by angle θ." We mentioned previously that matrices are really functions – this is a case of that fact.

When $\theta = 0$, we are rotating by an angle 0; in other words, we leave the points alone. Observe that $R(0) = I_2$, the 2×2 identity matrix, and so equation (3.3), in this case, confirms the signal property of the identity matrix.

The rotation matrices obey an important identity: for all α and β we have $R(\alpha) \cdot R(\beta) = R(\alpha + \beta)$. This can be seen in at least two ways. The first is to multiply out $R(\alpha) \cdot R(\beta)$ and use the angle addition formulas to simplify the entries – if you do this, you will see that you get the entries of $R(\alpha + \beta)$. Well and good. But there is another way to see that these matrices are equal, and this second way hints at a profound idea, namely that matrix multiplication is really function composition!

---

[2]If you wish to verify the formula for the rotated point, write (x, y) is polar coordinates $(s \cdot \cos(\alpha), s \cdot \sin(\alpha))$, so that rotation by angle θ moves (x, y) to $(s \cdot \cos(\alpha + \theta), s \cdot \sin(\alpha + \theta))$ and use the angle addition formulas.

Suppose we rotate by an angle β, and then rotate by an angle α. We can consider this to be one rotation after another, the way it was just described, but it is also the same as rotating by $\alpha + \beta$ to begin with. Taking the "angle by angle" approach, rotating (x, y) by angle β is multiplication by $R(\beta)$. Rotating the resulting point

$$R(\beta) \cdot \begin{pmatrix} x \\ y \end{pmatrix}$$

by angle α is multiplication by $R(\alpha)$:

$$R(\alpha) \cdot \left(R(\beta) \cdot \begin{pmatrix} x \\ y \end{pmatrix} \right)$$

Now the lowly associative law comes into play.

$$R(\alpha) \cdot \left(R(\beta) \cdot \begin{pmatrix} x \\ y \end{pmatrix} \right) = \left(R(\alpha) \cdot R(\beta) \right) \cdot \begin{pmatrix} x \\ y \end{pmatrix}$$

On the other hand, rotating by $\alpha + \beta$ all at once is multiplication by $R(\alpha + \beta)$:

$$R(\alpha + \beta) \cdot \begin{pmatrix} x \\ y \end{pmatrix}$$

Putting the two approaches together, we get

$$(3.4) \qquad R(\alpha + \beta) \cdot \begin{pmatrix} x \\ y \end{pmatrix} = \left(R(\alpha) \cdot R(\beta) \right) \cdot \begin{pmatrix} x \\ y \end{pmatrix}$$

We want to cancel the (x, y) from each side of the equation in (3.4). Unfortunately, there is no general cancellation law for matrices. However, the fact that equation (3.4) holds *for all* (x, y), rather than just for some particular (x, y), shows that it can be cancelled in this case. Indeed, we can let (x, y) be the first column of I_2, and Proposition 3.1 shows that

$$R(\alpha + \beta) \cdot \begin{pmatrix} 1 \\ 0 \end{pmatrix} = \text{first column of } R(\alpha + \beta)$$

and it shows that

$$\left(R(\alpha) \cdot R(\beta) \right) \cdot \begin{pmatrix} 1 \\ 0 \end{pmatrix} = \text{first column of } R(\alpha) \cdot R(\beta)$$

Equation (3.4) shows that the first column of $R(\alpha + \beta)$ is equal to the first column of $R(\alpha) \cdot R(\beta)$. A similar argument can be made for the second columns, and we have

$$R(\alpha + \beta) = R(\alpha) \cdot R(\beta)$$

Thus, rotation matrix multiplication reflects that "rotate by β" with "rotate by α" is the composite function "rotate by $\alpha + \beta$."

3. Problems

3.1. Prove the associative law of matrix addition.

3.2. Show that scalar multiplication distributes over matrix addition: that

$$\alpha \cdot (A + B) = (\alpha \cdot A) + (\alpha \cdot B)$$

for all $m \times n$ matrices A, B and numbers α.

3.3. Compute $A \cdot B$ in each of the following. (Note: in (b), be careful about the size of the result!)

a) $A = \begin{pmatrix} 0 & -1 & 2 \\ 4 & 5 & 6 \\ 1 & 0 & 1 \end{pmatrix}$ $B = \begin{pmatrix} 3 & -1 \\ 10 & 1/2 \\ 0 & 2 \end{pmatrix}$ **b)** $A = \begin{pmatrix} -2 \\ 3 \\ 6 \end{pmatrix}$ $B = \begin{pmatrix} -1 & 0 & 2 & 4 \end{pmatrix}$

c) $A = R(1 + \pi/3)$, $B = R(\pi/6 - 1)$ **d)** $A = \begin{pmatrix} 1 & 0 \\ 2 & 3 \end{pmatrix}$ $B = \begin{pmatrix} 0 & 2 \\ 2 & 0 \end{pmatrix}$

3.4. Find all matrices A such that

$$A \cdot \begin{pmatrix} 2 & 3 \\ 1 & 1 \end{pmatrix} = \begin{pmatrix} -4 & 0 \\ 1 & 1 \\ 0 & 2 \end{pmatrix}$$

3.5. Let b, c be arbitrary numbers with $c \neq 0$ and define

$$A = \begin{bmatrix} b & c \\ (1 - b^2)/c & -b \end{bmatrix}$$

Show that $A^2 = I_2$. (Note: thus, there are infinitely many *square roots of 1* among the matrices.)

3.6. Show that $I_m \cdot A = A$ for all $m \times n$ matrices A.

3.7. Let A be $m \times n$ and let C be $n \times r$. Let j be an integer with $1 \leq j \leq r$. Show that the j-th column of $A \cdot C$ is A times (the j-th column of C).

3.8. In each of the following two cases, compute A^2, A^3, and so on until you see a pattern. What is A^k for an arbitrary positive integer k in each case?

$$A = \begin{pmatrix} 0 & -1 & 0 \\ 0 & 0 & 1 \\ 1 & 0 & 0 \end{pmatrix} \qquad\qquad A = \begin{pmatrix} 1 & 2 \\ 0 & 1 \end{pmatrix}$$

3.9. Find a 2×1 matrix u *with non-zero complex number entries*, such that $u^T \cdot u = 0$.

3.10. Find a non-zero matrix A such that

$$A \cdot \begin{bmatrix} 1 & 2 \\ 3 & 4 \\ 5 & 6 \end{bmatrix} = \begin{bmatrix} 0 & 0 \\ 0 & 0 \end{bmatrix}$$

Show that there is **no** non-zero matrix B such that

$$B \cdot \begin{bmatrix} 1 & 2 \\ 3 & 4 \end{bmatrix} = \begin{bmatrix} 0 & 0 \\ 0 & 0 \end{bmatrix}$$

3.11. For each real number c, define

$$f(c) = \begin{bmatrix} c & 0 \\ 0 & c \end{bmatrix}$$

For real numbers c, d, show that $f(c+d) = f(c) + f(d)$ and $f(c \cdot d) = f(c) \cdot f(d)$. Define $i = R(\pi/2)$, and show that $i^2 = f(-1)$. Does this suggest a way to represent the complex numbers using matrices?

3.12.   (Continuing the previous problem.) Define

$$r = \begin{bmatrix} 0 & 1 \\ 2 & 0 \end{bmatrix}$$

What real number corresponds to r^2? Use matrices to show that

$$(\sqrt{2} - 1) \cdot (\sqrt{2} + 1) = 1$$

CHAPTER 4

Linear Equations

1. Equations and Solutions

One of the main uses of matrices is to represent linear equations. A *linear equation* in variables $x_1, x_2, \ldots x_n$ looks like

$$(4.1) \qquad a_1 x_1 + a_2 x_2 + \cdots + a_n x_n = b$$

where the a_i and b are constants. Observe that the left side of this last equation is a *dot product*, that is to say, it is a matrix multiplication.

$$\begin{pmatrix} a_1 & a_2 & \cdots & a_n \end{pmatrix} \cdot \begin{pmatrix} x_1 \\ x_2 \\ \vdots \\ x_n \end{pmatrix} = b$$

Thus, a linear equation can be written as a $1 \times n$ matrix of coefficients times an $n \times 1$ matrix of variables, set equal to a constant.

A *solution* to (4.1) is a set of values of the variables:

$$x_1 = c_1, \ x_2 = c_2, \ \ldots, \ x_n = c_n$$

that makes the equation true:

$$a_1 c_1 + \cdots + a_n c_n = b$$

In matrix terms, we have

$$\begin{pmatrix} a_1 & a_2 & \cdots & a_n \end{pmatrix} \cdot \begin{pmatrix} c_1 \\ c_2 \\ \vdots \\ c_n \end{pmatrix} = b$$

A *system of linear equations* is a finite set of linear equations, all in the same variables. We might have variables $x_1, \ldots, x_n$, and imagine, say, m equations in them. We will use a suggestive matrix notation for the coefficients: $A[i,j]$ will be the coefficient in the i-th equation of the variable x_j. Here is our system.

$$
\begin{array}{ccccccccc}
A[1,1]x_1 & + & A[1,2]x_2 & + & \cdots & + & A[1,n]x_n & = & b_1 \\
A[2,1]x_1 & + & A[2,2]x_2 & + & \cdots & + & A[2,n]x_n & = & b_2 \\
\vdots & & \vdots & & \vdots & & \vdots & & \vdots \\
A[m,1]x_1 & + & A[m,2]x_2 & + & \cdots & + & A[m,n]x_n & = & b_m
\end{array}
$$

(4 2)

A *solution* to the system (4.2) is a simultaneous solution to each of the equations in the system: specific values of each of $x_1, x_2, \ldots, x_n$ that make each of the m equations true.

The matrix product version of each linear equation leads directly to a matrix representation of the entire system (4.2). The coefficients have already been indexed to live in an $m \times n$ matrix A; let X be the $n \times 1$ matrix of the variables, so that $x_j = X[j,1]$ for each j, and let B be the $m \times 1$ *right side matrix,* so that $B[i,1] = b_i$ for each i. Then (4.2) is

$$(4.3) \qquad\qquad A \cdot X = B$$

The matrix A is the *coefficient matrix* of the system; the matrix B is the *right side.* The number of rows of A is the number of equations in the system, and the number of columns of A is the number of variables in the system. If A_i is the i-th row of A, then the i-th equation in the system (4.3) is $A_i X = B[i,1]$. A solution to the system (4.3) is an $n \times 1$ matrix C such that

$$A \cdot C = B$$

The entry $C[j,1]$ in C tells the value of the variable $X[j,1] = x_j$. Taking an "equation by equation" approach, that C is a solution to the i-th equation for each i says that $A_i C = B[i,1]$ for each i.

It is not hard to solve a system of linear equations. The main idea is to use one of the equations to solve for one of the variables; then you can substitute for that variable in the remaining equations, reducing the problem by one equation and one variable, continuing in the same way. We could take a rather ad-hoc approach to the solution of linear equations, but we choose to be systematic for at least three reasons. First, there is numerical information hidden in the equation $AX = B$ that is relevant to general applications of matrices; a careful solution technique will disclose this information. Second, a systematic approach to solutions will allow us to reach a solution in an efficient way that avoids dead ends. Third, an algorithmic solution technique can be programmed! Indeed, this algorithm is incorporated into a wide variety of computer software. We will discuss this in class to some extent, but you should know in advance that the programming involved is not trivial because round-off errors in calculations can produce meaningless results *even if the round-off is extremely small.*

2. Elimination: Free and Pivoted Unknowns

The algorithm we will use is called Gauss-Jordan Elimination. It is a simple algorithm whose name does not need two famous mathematicians, and therefore we will call it simply "Elimination." Everything we do with matrices and with systems of equations will be understood by having a thorough understanding of Elimination. You should be elated that the entire subject of linear algebra can be understood by mastering what we have called a "simple algorithm." On the other hand, if you have come to expect the subtleties that attend every mathematical subject, you will make sure you really do develop a thorough understanding of Elimination. The linear algebra we will build from it is not obvious.

Elimination involves three types of operations that can be performed on a matrix. These operations are the *elementary operations*.[1]

(a) Interchange two of the rows.

(b) Multiply a row by a non-zero number.

(c) Add a multiple of one row to another (leaving the first row the same).

We need two facts about elementary operations. The first is that each operation has an *inverse operation* that undoes what the operation does. To see this, suppose first that we switch rows i and j in the matrix A to obtain the matrix A'. Then switching rows i and j on A' brings us back to A. The elementary operation "switch rows i and j" has the inverse operation "switch rows i and j." Suppose next that we multiply row i of A by $\alpha \neq 0$ to obtain A'. Then multiplying row i of A' by $1/\alpha$ gives us A again. Thus, the elementary operation "multiply row i by α" has inverse operation "multiply row i by $1/\alpha$."

Finally, suppose that we add α times row i of A to row j of A, thereby producing the matrix A'. (Remember, we don't change row i.) If we add $-\alpha$ times row i of A' to row j of A', we will obtain A. Thus, the operation "add α times row i to row j" has inverse "add $-\alpha$ times row i to row j."

Our second fact begins with the equation $A \cdot X = B$ and applies an elementary operation to A, B (the same operation to both) to get $A' \cdot X = B'$. This kind of manipulation will furnish the guts of Elimination. We claim that if C is a solution to $A \cdot X = B$, then C is a solution to $A' \cdot X = B'$. To see this, we need to remember that the i-th equation in $A \cdot X = B$ corresponds to the i-th row A_i of A: that equation is $A_i \cdot X = B[i]$. So, if the operation is to switch rows i and j, then the *set* of equations doesn't change – just their position. Obviously, that doesn't change the set of solutions. If the operation multiplies row i by α, then the i-th equation $A_i \cdot X = B[i]$ becomes $\alpha \cdot A_i \cdot X = \alpha \cdot B[i]$, and if C solves the first equation, it solves the second. If the operation adds

---

[1]These operations are also called *row operations* and *elementary row operations*.

α times row i to row j, then the i-th equation stays the same and the j-th changes from $A_j \cdot X = B[j]$ to

$$(\alpha \cdot A_i + A_j) \cdot X = \alpha \cdot B[i] + B[j]$$

If $A_i \cdot C = B[i]$ and $A_j \cdot C = B[j]$, then

$$(\alpha \cdot A_i + A_j) \cdot C = \alpha \cdot A_i \cdot C + A_j \cdot C = \alpha \cdot B[i] + B[j]$$

and C is a solution to the transformed equation. Thus, if C is a solution to $A \cdot X = B$, it is a solution to $A' \cdot X = B'$ where the latter equation is obtained by applying an elementary operation.

We put two of the previous ideas together to show that $A \cdot X = B$ and $A' \cdot X = B'$ have *exactly the same solutions*. This is because there is an inverse operation that changes A', B' back into A, B. So, if $A' \cdot C = B'$, then $A \cdot C = B$, applying the reasoning of the previous paragraph. We see that elementary operations on $A \cdot X = B$ do not change the set of solutions.

Elimination will apply elementary operations to transform a given system into a form in which the solutions (or lack of solutions) will be obvious. Since the operations do not change the set of solutions, the solutions of the given system will be obvious from the form of the transformed system.

We will describe Elimination abstractly, at the same time working with a specific example. Given the system $AX = B$, for instance

$$\begin{pmatrix} 1 & 2 & 3 \\ 4 & 5 & 6 \\ 7 & 8 & 9 \end{pmatrix} \begin{pmatrix} x_1 \\ x_2 \\ x_3 \end{pmatrix} = \begin{pmatrix} 1 \\ 0 \\ 0 \end{pmatrix}$$

Elimination puts the coefficient matrix A with the right side matrix B to form a single $m \times (n+1)$ matrix $[A|B]$. (The vertical line is just meant to separate the A from the B so that we do not confuse this juxtaposition with the matrix product AB.)

$$[A|B] = \begin{pmatrix} 1 & 2 & 3 & 1 \\ 4 & 5 & 6 & 0 \\ 7 & 8 & 9 & 0 \end{pmatrix}$$

Since B is $m \times 1$, it fits in next to the columns of A. The matrix $[A|B]$ is the *augmented matrix* of the system. The augmented matrix is just a table that contains *all* the numbers we are working with.

Here is the Gauss-Jordan Elimination algorithm.

Elimination

Apply Steps 1-4, with row $1, 2, \ldots$, in turn, as *current row*, until the bottom row is reached or Step 1 fails.

Step 1. Find the leftmost column in the coefficient part of the augmented matrix having a non-zero entry at the current row or below. If there is no such entry, Step 1 fails. Otherwise, choose one such entry (this entry is a *pivot*).

Step 2. Switch rows, if necessary, to bring the pivot to the current row.

Step 3. Multiply the current row by the inverse of the pivot (so that the pivot now has value 1).

Step 4 Add multiples of the current row to rows above and below it so that the pivot is the only non-zero entry in its column. ■

In trying the algorithm on our example, observe the notation we use for the elementary operations. Rows are referred to by roman numerals. We use $-(1/3) \cdot \text{II}$ for multiplying row 2 by -1/3, and $-4 \cdot \text{I} + \text{II}$ for adding -4 times row 1 to row 2. Here is the augmented matrix.

(4.4) $$\begin{pmatrix} 1 & 2 & 3 & 1 \\ 4 & 5 & 6 & 0 \\ 7 & 8 & 9 & 0 \end{pmatrix}$$

We start with row 1 as the *current row*. The leftmost column having a non-zero entry at row 1 or below is column 1. We use the 1,1 entry 1 as pivot.

Step 2 and Step 3 are unnecessary; here are the calculations for Step 4. Notice that the row with the pivot (row 1) is not changed.

$$\begin{pmatrix} 1 & 2 & 3 & 1 \\ 4 & 5 & 6 & 0 \\ 7 & 8 & 9 & 0 \end{pmatrix} \quad \begin{matrix} -4 \cdot \text{I} + \text{II} \\ -7 \cdot \text{I} + \text{III} \end{matrix} \quad \begin{pmatrix} 1 & 2 & 3 & 1 \\ 0 & -3 & -6 & -4 \\ 0 & -6 & -12 & -7 \end{pmatrix}$$

Now we use row 2 as the *current row*. In rows 2 and below, the leftmost non-zero entries are in column 2, we choose -3 as pivot. Step 2 is unnecessary; Step 3 multiplies row 2 by $-1/3$, so that the transformed entry at $2,2$ is 1. Then Step 4 clears the entries in column 2 above and below the pivot. Since the $1,2$-entry is 2, we add -2 times row 2 to row 1 to clear the $1,2$-entry.

$$(4.5) \quad \begin{pmatrix} 1 & 2 & 3 & 1 \\ 0 & -3 & -6 & -4 \\ 0 & -6 & -12 & -7 \end{pmatrix} \quad \begin{matrix} -(1/3) \cdot \text{II} \\ -2 \cdot \text{II} + \text{I} \\ 6 \cdot \text{II} + \text{III} \end{matrix} \quad \begin{pmatrix} 1 & 0 & -1 & -5/3 \\ 0 & 1 & 2 & 4/3 \\ 0 & 0 & 0 & 1 \end{pmatrix}$$

Now row 3 is the current row. There are no non-zero entries at row 3 or below (we do not count the 1 at entry 1,4, since that entry is not in the coefficient matrix). Step 1 fails and the algorithm stops. The coefficient part of the final augmented matrix obtained is said to be in *row-echelon form* to indicate the placement of pivots left to right and top to bottom in their "echelons."

We showed that elementary operations do not change the set of solutions. Thus, the solutions to (4.5) are the same as those for the original system (4.4). But the last equation in (4.5) says

$$0x_1 + 0x_2 + 0x_3 = 1$$

and this is impossible. Thus, the system (4.4) has no solutions at all. We say the system is *inconsistent*. This example shows the general form of an inconsistent system at the end of Elimination. There will be a row of zeros in the coefficient matrix and a non-zero right side. It is clear that such an equation cannot have solutions. We will see momentarily that if a system is inconsistent, we always arrive at this form.

Another example. Here is the augmented matrix.

$$\begin{pmatrix} 2 & -3 & 1 & 4 & 1 & 17 \\ -4 & 6 & -1 & -6 & 0 & -27 \\ 1 & 1 & 2 & 7 & -1 & 20 \\ -4 & 1 & -4 & -16 & 3 & -50 \end{pmatrix}$$

Step 1 might choose the 2 at the 1,1-entry as pivot. Step 2 is not needed, and Step 3 and Step 4 compute

$$\begin{pmatrix} 2 & -3 & 1 & 4 & 1 & 17 \\ -4 & 6 & -1 & -6 & 0 & -27 \\ 1 & 1 & 2 & 7 & -1 & 20 \\ -4 & 1 & -4 & -16 & 3 & -50 \end{pmatrix} \begin{matrix} (1/2)\cdot\text{I} \\ 4\cdot\text{I}+\text{II} \\ -1\cdot\text{I}+\text{III} \\ 4\cdot\text{I}+\text{IV} \end{matrix} \begin{pmatrix} 1 & -3/2 & 1/2 & 2 & 1/2 & 17/2 \\ 0 & 0 & 1 & 2 & 2 & 7 \\ 0 & 5/2 & 3/2 & 5 & -3/2 & 23/2 \\ 0 & -5 & -2 & -8 & 5 & -16 \end{pmatrix}$$

With row 2 as current row, the leftmost column at row 2 or below is column 2. We use $5/2$ as pivot; Step 2 interchanges rows 2 and 3 (notice how this is denoted):

$$\begin{pmatrix} 1 & -3/2 & 1/2 & 2 & 1/2 & 17/2 \\ 0 & 0 & 1 & 2 & 2 & 7 \\ 0 & 5/2 & 3/2 & 5 & -3/2 & 23/2 \\ 0 & -5 & -2 & -8 & 5 & -16 \end{pmatrix} \text{II} \leftrightarrow \text{III} \begin{pmatrix} 1 & -3/2 & 1/2 & 2 & 1/2 & 17/2 \\ 0 & 5/2 & 3/2 & 5 & -3/2 & 23/2 \\ 0 & 0 & 1 & 2 & 2 & 7 \\ 0 & -5 & -2 & -8 & 5 & -16 \end{pmatrix}$$

Then Step 3 and Step 4 come along.

$$\begin{pmatrix} 1 & -3/2 & 1/2 & 2 & 1/2 & 17/2 \\ 0 & 5/2 & 3/2 & 5 & -3/2 & 23/2 \\ 0 & 0 & 1 & 2 & 2 & 7 \\ 0 & -5 & -2 & -8 & 5 & -16 \end{pmatrix} \begin{matrix} (2/5)\cdot\text{II} \\ (3/2)\cdot\text{II}+\text{I} \\ 5\cdot\text{II}+\text{IV} \end{matrix} \begin{pmatrix} 1 & 0 & 7/5 & 5 & -2/5 & 77/5 \\ 0 & 1 & 3/5 & 2 & -3/5 & 23/5 \\ 0 & 0 & 1 & 2 & 2 & 7 \\ 0 & 0 & 1 & 2 & 2 & 7 \end{pmatrix}$$

Now row 3 is the current row. We use the 1 at entry 3,3 as pivot. Step 2 and Step 3 are skipped. Step 4:

$$\begin{pmatrix} 1 & 0 & 7/5 & 5 & -2/5 & 77/5 \\ 0 & 1 & 3/5 & 2 & -3/5 & 23/5 \\ 0 & 0 & 1 & 2 & 2 & 7 \\ 0 & 0 & 1 & 2 & 2 & 7 \end{pmatrix} \begin{matrix} -(3/5)\cdot\text{III}+\text{II} \\ -(7/5)\cdot\text{III}+\text{I} \\ -1\cdot\text{III}+\text{IV} \end{matrix} \begin{pmatrix} 1 & 0 & 0 & 11/5 & -16/5 & 28/5 \\ 0 & 1 & 0 & 4/5 & -9/5 & 2/5 \\ 0 & 0 & 1 & 2 & 2 & 7 \\ 0 & 0 & 0 & 0 & 0 & 0 \end{pmatrix}$$

Continue to row 4 and Step 1 fails. We have row echelon form.

The final system has a row of coefficient 0's, just as occurred in the previous example, but this time the right side is 0. The last equation in this system simply says $0 = 0$, which is always true and, therefore, has no effect on solutions. Let's write out the three equations that matter.

$$x_1 + (11/5)x_4 - (16/5)x_5 = 28/5$$
$$x_2 + (4/5)x_4 - (9/5)x_5 = 2/5$$
$$x_3 + 2x_4 + 2x_5 = 7$$

If x_4 and x_5 are chosen arbitrarily, then x_1 and x_2 and x_3 are uniquely determined in a solution to the system. Thus, there are infinitely many solutions, and, in choosing some particular solution, x_4 and x_5 are arbitrary. These arbitrary variables did not get pivots in Elimination, whereas the determined variables did get pivots. When a system is consistent (has solutions), the pivoted variables are determined by the non-pivoted variables. The non-pivoted variables are said to be *free* since their values are arbitrary. The presence of free variables is what signals that there are infinitely many solutions.

We have remarked that the pivoted equations form a consistent system (with one solution for each choice of free variables). It follows that the only way for a system to be inconsistent is to have a non-pivoted row (a row of zeros in the coefficient matrix). If the right side across from this row is 0, then the equation is 0=0 which is redundant. Therefore, to be inconsistent, we must have a row of zeros in the row-echelon coefficient matrix across from a non-zero right side. The first system we solved had this property, and now we see that this is the only way for a system to be inconsistent.

Step 1 asks us to choose an appropriate matrix entry for a pivot. We might ask about the effect of different allowed choices; we will have definitive information later in the course.

The pivots are chosen from the coefficient matrix without looking at the right side entries. The number of pivots (pivoted variables) in Elimination is called the *rank* of the coefficient matrix. The number of free variables (non-pivoted variables) is the *nullity* of the coefficient matrix. These numbers are very important, and they tell us a lot about a given system. The sum of the rank and nullity is the number of variables, and this is the number of columns in the coefficient matrix. Later we will prove that the rank and nullity do not depend on any choices made in Step 1.

A third example, with augmented matrix

$$\begin{pmatrix} -1 & 0 & 1 & 1 \\ 2 & 0 & 2 & 2 \\ 1 & 3 & 3 & 3 \end{pmatrix}$$

We will show the steps in the computation, leaving it to you to observe which pivots are used and what operations need to be done.

$$\begin{pmatrix} -1 & 0 & 1 & 1 \\ 2 & 0 & 2 & 2 \\ 1 & 3 & 3 & 3 \end{pmatrix} \implies \begin{pmatrix} 1 & 0 & -1 & -1 \\ 0 & 0 & 4 & 4 \\ 0 & 3 & 4 & 4 \end{pmatrix}$$

$$\begin{pmatrix} 1 & 0 & -1 & -1 \\ 0 & 0 & 4 & 4 \\ 0 & 3 & 4 & 4 \end{pmatrix} \implies \begin{pmatrix} 1 & 0 & -1 & -1 \\ 0 & 3 & 4 & 4 \\ 0 & 0 & 4 & 4 \end{pmatrix} \implies \begin{pmatrix} 1 & 0 & -1 & -1 \\ 0 & 1 & 4/3 & 4/3 \\ 0 & 0 & 1 & 1 \end{pmatrix}$$

$$\begin{pmatrix} 1 & 0 & -1 & -1 \\ 0 & 1 & 4/3 & 4/3 \\ 0 & 0 & 1 & 1 \end{pmatrix} \implies \begin{pmatrix} 1 & 0 & 0 & 0 \\ 0 & 1 & 0 & 0 \\ 0 & 0 & 1 & 1 \end{pmatrix}$$

What is the rank? What is the nullity? Is the system consistent? How many solutions are there?

We turn to a fairly special situation in order to make an observation that we will need later on. Let A be an $n \times n$ coefficient matrix, and suppose that A has rank n. We claim that the row-echelon form of A is the identity matrix I_n. The reason for this? The first pivot goes in row 1, the second in row 2, and so on. Since the rank is n, there are n pivots, and therefore each of the n rows must get a pivot. The pivots go in distinct columns, left to right,[2] and since A has n pivots, each column must get a pivot, left to right. Thus, the pattern of pivots is along the diagonal (top to bottom, left to right) of A. Because Step 3 divides by each pivot, the diagonal positions in row-echelon form are 1's. Because we eliminate above and below each pivot, all the non-diagonal positions in A are 0's. Thus, the row-echelon form is I_n.

Again suppose that A is $n \times n$ of rank n, and suppose we use Elimination to solve the system $AX = B$. We begin with the augmented matrix $[A|B]$, and we end up with row-echelon form $[I_n|B']$, where the first n columns are those of the identity matrix, as we proved the last paragraph. This linear equation is $I_n \cdot X = B'$, and $X = B'$ is obviously the unique solution! Thus, $A \cdot B' = B$, and B' is the only solution to $AX = B$. We will use this idea in the next section.

Returning to the subject of Elimination in general, because the pivots are chosen from the coefficient matrix (not from the right side column), if we have two systems with the same coefficient matrix, we can solve both systems with the same Elimination steps. For instance, suppose we want to solve

$$\begin{pmatrix} 1 & -1 & 2 \\ 2 & 1 & 3 \\ 4 & 5 & 5 \end{pmatrix} X = \begin{pmatrix} 1 \\ 4 \\ 3 \end{pmatrix} \quad \text{and} \quad \begin{pmatrix} 1 & -1 & 2 \\ 2 & 1 & 3 \\ 4 & 5 & 5 \end{pmatrix} X = \begin{pmatrix} 2 \\ 7 \\ 17 \end{pmatrix}$$

We augment the coefficient matrix by both right sides

$$\begin{pmatrix} 1 & -1 & 2 & 1 & 2 \\ 2 & 1 & 3 & 4 & 7 \\ 4 & 5 & 5 & 3 & 17 \end{pmatrix}$$

---

[2]Here is a proof, just to be above-board! Once a pivot is found, say in row r, Step 4 clears the column below the pivot. We never again look for pivots at row r or above. Furthermore, the pivot was found in the leftmost column having a non-zero entry. Thus, when Step 1 looks for pivots, the columns at and to the left of the previously found pivots are all 0. Any new pivot will therefore be to the right of the old ones.

and use Elimination exactly as stated, choosing pivots from the coefficient part of the augmented matrix (from the first 3 columns). Check that Elimination yields

$$\begin{pmatrix} 1 & 0 & 5/3 & 5/3 & 3 \\ 0 & 1 & -1/3 & 2/3 & 1 \\ 0 & 0 & 0 & -7 & 0 \end{pmatrix}$$

The first three columns of this last matrix express the coefficients in row-echelon form. The rank is 2 and the nullity is 1. We can see that the first system is inconsistent:

$$\begin{pmatrix} 1 & 0 & 5/3 & 5/3 \\ 0 & 1 & -1/3 & 2/3 \\ 0 & 0 & 0 & -7 \end{pmatrix}$$

The second system has infinitely many solutions:

$$\begin{pmatrix} 1 & 0 & 5/3 & 3 \\ 0 & 1 & -1/3 & 1 \\ 0 & 0 & 0 & 0 \end{pmatrix} \quad \text{yields} \quad \begin{aligned} x_1 &= 3 - (5/3)x_3 \\ x_2 &= 1 + (1/3)x_3 \\ x_3 &\text{ is arbitrary (free).} \end{aligned}$$

We can also do Elimination on a coefficient matrix with no right side at all. This will not solve a system of equations, but it does tell us the rank and nullity of the matrix, and we will see that these numbers are important beyond their use in linear equations.

To recap, the rank is the number of pivots encountered in Elimination, the nullity is the number of free variables (the variables that do not get pivots). The rank plus the nullity is the number of variables, and this is the number of columns of the coefficient matrix. After Elimination, if there is a row of 0's in the coefficient matrix and a non-zero right side across from that row, then the system is inconsistent, having no solutions. If there is no such row, then the system is consistent. After discarding the rows of 0's, each row gives the value of a pivoted variable in terms of free variables. There is a solution for each arbitrary choice of free variables. If there are no free variables in this case, then the system has a unique solution.

3. Problems

4.1. Use Elimination (by hand) to solve the systems of equations, noting the pivots and elementary operations along the way and remarking the rank and nullity at the end.

a) $\begin{aligned} 2A - 2B - 10C &= 4 \\ -2A + 3B + 13C &= \text{-3} \\ 3B + 9C &= 3 \end{aligned}$
b) $\begin{aligned} x_1 - 2x_2 + 2x_3 - 3x_4 &= 19 \\ -3x_1 + 6x_2 - 8x_3 + 13x_4 &= \text{-71} \end{aligned}$

c) $\begin{aligned} p - 2q - r &= \text{-6} \\ 2p - 2q - 2r &= \text{-8} \\ -p + q + 2r &= 3 \\ -p + 2q + 2r &= 5 \end{aligned}$
d) $\begin{aligned} x + z + w &= 2 \\ x - y + 3z - 2w &= \text{-3} \\ 4x - 3y + 10z - 5w &= \text{-5} \end{aligned}$

4.2. Find b so that this system is consistent.

$$x_1 - 3x_2 + 5x_3 = 4$$
$$2x_1 + x_2 - 3x_3 = b$$
$$5x_1 + 6x_2 - 14x_3 = 8$$

4.3. Find all matrices A such that

$$A \cdot \begin{pmatrix} 2 & 1 \\ -3 & 1 \end{pmatrix} = \begin{pmatrix} 2 & 1 \\ -3 & 1 \end{pmatrix} \cdot A$$

(Hint: what is the size of A? Let its entries be unknowns and solve equations.)

4.4. Use Elimination to solve the following system of equations (given in terms of its augmented matrix), but perform the arithmetic steps to three significant digits with an exact base-10 exponent. Find the approximate solution this way. Now solve the system exactly. Is there much difference between the approximate and exact solutions?

$$\begin{pmatrix} 5 & 5 \cdot 10^5 & 6.18 \cdot 10^5 \\ 2 \cdot 10^{-5} & 3 & 3.72 \end{pmatrix}$$

4.5. In each case, find an example of a system of linear equations with the indicated features. (Feel free to use examples from other homework problems or from class.)

a) Same number of equations as variables, infinitely many solutions.

b) Less equations than variables, no solution.

c) Less variables than equations, unique solution.

4.6. When Elimination is performed on the 2×3 matrix A (as coefficient matrix with no right side), what are the possible row-echelon forms that could result? (Use 1's and 0's where they have to occur; use stars for unknown entries.)

Application problems. The following will be discussed in class, using specific examples and more general theory.

4.7. (*Leontief's model of an economy*) We have an economy consisting of n people, each of whom produces one unit of a unique product. Person i produces $P[i]$ worth of product. (You can think of $P[i]$ as i's income.) Person i purchases the fraction $A[i, j]$ of the product made by person j. (So that $0 \le A[i, j] \le 1$.) We assume that all of what is produced is purchased. What does this say about the matrix A? What does the matrix $A \cdot P$ measure? Why might it be interesting to know whether there is an $n \times 1$ matrix P such that $A \cdot P = P$? (The existence of P is that the economy is *closed*.)

4.8. The table below gives the matrix A of the previous problem in a Leontief model. Solve for the price vector P.

| | Units purchased each year | | | |
	wheat	bread	wood	tables
wheat producer	0	0.3	0.15	0.2
bread producer	0.8	0.2	0.2	0.3
wood producer	0.1	0.25	0.1	0.4
tables producer	0.1	0.25	0.55	0.1
unit price	$P[1]$	$P[2]$	$P[3]$	$P[4]$

4.9. (*Linear Markov process*) Suppose that if it is sunny today, then there is a definite chance p (so that $0 \le p \le 1$) that it will be sunny tomorrow. Suppose that if it is not sunny, it is rainy, so there the chance of rain the day after a sunny day is $1 - p$. Suppose that if it is rainy today, the chance it is sunny tomorrow is q, and the chance it is rainy tomorrow is $1 - q$. Today it is sunny. What is your forecast for 10 days from now? What happens after a long period of time?

4.10. (*Polynomial values.*) Given arbitrary coefficients $c_0, c_1, \ldots, c_n$, we can define a polynomial

$$f(x) = c_0 + c_1 \cdot x + \cdots + c_n \cdot x^n$$

We say this polynomial has degree *at most* n, since c_n could be 0. We want to notice that $f(x)$ is a matrix product:

$$f(x) = \begin{pmatrix} 1 & x & x^2 & \cdots & x^n \end{pmatrix} \cdot \begin{pmatrix} c_0 \\ c_1 \\ \vdots \\ c_n \end{pmatrix}$$

We will see that it will be useful to regard the right column as *variable*. Call it C. Now suppose we have m values $x_1, x_2, \ldots, x_m$. Express the column

$$\begin{pmatrix} f(x_1) \\ f(x_2) \\ \vdots \\ f(x_m) \end{pmatrix}$$

as $V \cdot C$ where V is $m \times (n + 1)$. (The matrix V is called a *Vandermonde matrix*[3] for $x_1, \ldots, x_m$.)

---

[3]That Vandermonde's name is attached to this matrix is apparently a mistake! In the paper [**9**], the mathematician Lebesgue claims that the mistake is due to a misreading of the notation used by Vandermonde in one of his papers.

4.11. (Continuation of the previous problem.) Suppose that $m \geq (n + 1)$, and let V be the $m \times (n+1)$ Vandermonde matrix for $x_1, \dots, x_m$ where the x_j are distinct. Show that $V \cdot X = \mathbb{O}_{m \times 1}$ has a unique solution. (Hint: remember the algebraic fact: a non-zero polynomial of degree at most n can have at most n roots. If $V \cdot C = \mathbb{O}$ with $C \neq \mathbb{O}$, then the polynomial $f(x)$ formed from C is a non-zero polynomial; what are its roots?)

4.12. Find a polynomial $f(x)$ of degree at most 2 such that $f(2) = 3$ and $f(3) = 10$ and $f'(2) = 7$.

4.13. (*Kirchoff's Laws.*) The word *graph* is often used for a set of *vertices* (points), some of which are connected by *edges* (curves). A simple electrical circuit is a graph in which the vertices are *junctions* and the edges are connecting wires or components. In each edge there is a *current*[4] J. If we choose a *direction* for each edge by putting an arrow on one end, then the sign of J indicates the direction of the current – with the arrow if $J > 0$ and against it if $J < 0$. *Kirchoff's Current Law*[5] asserts that, at each junction, the sum of the currents coming in (on arrows) is equal to the sum going out. Observe that this is a system of linear equations. A *tree* in a graph is a set of edges that does not contain a loop.[6] In the graph for an electrical circuit, choose a set of edges that forms a tree, using as many edges as possible. In the equation for Kirchoff's Current Law, this set of edges can be a set of pivoted variables. (The proof involves some elementary graph theory.) Demonstrate this in the case of the circuit here. Note that the currents have been labeled but you will need to choose a direction for each, and you will need to choose a maximal

---

[4]For those unfamiliar with electronics: current measures the number of electrons per unit time passing some point in the wire.

[5]This rule is also called *Kirchoff's Junction Rule*, and by other names as well. See [**16**].

[6]The idea of a loop is fairly intuitive; here is a formal definition: a *loop* is a list $v_1, \dots, v_k$ of vertices such that there is an edge between v_j and v_{j+1} for each j with $1 \leq j < k$. Also, $v_1 = v_k$ and $k \geq 2$. Thus, a loop travels over edges, ending up where it started. A loop does not have to follow the arrows.

tree. Choose two different trees and, in each case, show that the currents corresponding to edges in your tree can be used as pivots in Elimination. (Hint: in the columns of the augmented matrix, write the tree variables to the left.)

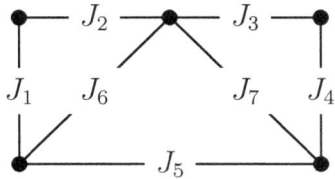

4.14. (Continuing the previous problem.) In a graph representing an electrical circuit, each edge has an associated *potential drop*[7] This *drop* can be positive or negative. There is a second Kirchoff's Law: the *Voltage Law*[8] which governs the potential: the sum of the drops around each loop is zero.[9] This is yet another system of linear equations! For each current J_i in the previous graph, define a potential V_i. The tree edges you found in the previous problem give the *free variables* V_i for the voltage law. Write down the equations for the Voltage Law and show that the tree edges can give the free variables. (Hint: write the non-tree edge variables to the left.)

4.15. Here is another circuit. Get the equations for the Current Law and for the Voltage Law. Think about basic variables in each case.

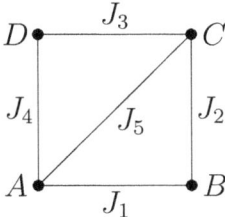

[7]Potential is usually measured in volts and represents electromotive force: work done per unit charge. The drop is often called the *voltage drop*.

[8]This rule is also called *Kirchoff's Loop Law*. See [**16**].

[9]As we travel around a loop, if potential V is encountered in the direction of an arrow, it counts as $+V$ around the loop; if we are moving against the arrow, then $-V$ is added.

4.16. The *Fibonacci numbers* form a sequence $F_0, F_1, \ldots$ defined by recursion: $F_0 = 1$, $F_1 = 1$, and $F_{n+2} = F_{n+1} + F_n$ for all $n \geq 0$. Show that there is a matrix M such that

$$\begin{pmatrix} F_{n+1} \\ F_{n+2} \end{pmatrix} = M \cdot \begin{pmatrix} F_n \\ F_{n+1} \end{pmatrix} \quad \text{for all} \quad n \geq 0$$

Now show that

$$\begin{pmatrix} F_n \\ F_{n+1} \end{pmatrix} = M^n \cdot \begin{pmatrix} F_0 \\ F_1 \end{pmatrix} \quad \text{for all} \quad n \geq 0$$

4.17. We have yellow flags and blue flags that are one foot tall, and we have red flags that are two feet tall. Let G_n be the number of ways to arrange yellow, blue, and red flags on n feet of flag pole. Explain why[10]

$$G_n = 2 \cdot G_{n-1} + G_{n-2} \quad \text{for} \quad n \geq 3$$

Also explain why $G_1 = 2$ and $G_2 = 3$. Find a matrix N such that

$$\begin{bmatrix} G_{n+1} \\ G_{n+2} \end{bmatrix} = N^n \cdot \begin{bmatrix} 2 \\ 3 \end{bmatrix} \quad \text{for} \quad n \geq 0$$

4.18. (*heat equilibrium*) In the graph below, the nodes are locations where heat is measured. The nodes labeled with single letters are kept at constant heat. The nodes labeled X_j can change as heat diffuses across the edges. We are interested in the equilibrium state where each X_j is constant. Diffusion dictates that each X_j is the average of the temperatures of nodes to which it is connected. Thus, for example,

$$X_1 = \frac{1}{3} \cdot \left[X_2 + A + B \right]$$

---

[10]This problem is a typical problem of *combinatorics* – the counting done in the course of that name.

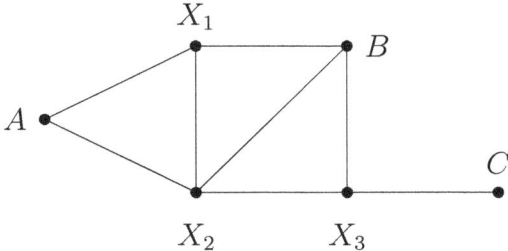

Show that the equilibrium equations have a unique solution, given that A, B, C are given constants. (Don't choose values for A, B, C; look at the rank of the coefficient matrix.) Now let $A = 5$ and $B = 2$ and $C = 20$ and solve for the X_j. (Suggestion: choose your pivots to avoid fractions.)

4.19. Another heat equilibrium problem. Show that if A, B are constants, there is always a unique solution for the X_j.

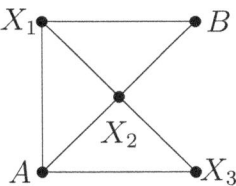

CHAPTER 5

Matrix Inverses

The simplest system of linear equations is a single equation in one variable; for numbers a, b the equation is

$$aX = b$$

Because such equations are so trivial, we can forget that there are actually three cases that can occur in them. First of all, if $a = 0$, then the equation is just $0 = b$, which probably seems silly, except that it can occur in practice (as it does sometimes when we do Elimination!). If $b \neq 0$, then $0 = b$ has *no solutions*, whereas if $b = 0$, then *every X is a solution*.

On the other hand, we usually have $aX = b$ with $a \neq 0$, and then we get the *unique* solution by division: $X = b/a$.

What will the foregoing look like when we have a system of linear equations $A \cdot X = B$? We ask whether we can *divide* by the matrix A, in some sense. Since matrix multiplication is not commutative, we have to be careful about notation. It is better to think in terms of an *inverse* than to try to invent matrix division. Going back to the simple equation $aX = b$, we could think of multiplying both sides by the inverse a^{-1}, rather than dividing by a, to get

$$a^{-1} \cdot a \cdot X = a^{-1} \cdot b \quad \text{which is} \quad X = a^{-1} \cdot b$$

We are going to do the same thing for *some* matrices. Remember that we can't divide by the number a unless it is not zero. For matrices, the situation is more complicated in that there are non-zero matrices that don't have an inverse. Let's begin with a formal definition and see where it leads.

The matrices A, C are *inverses* if $A \cdot C$ and $C \cdot A$ are both identity matrices. A matrix that has an inverse is said to be *invertible*. Another term is often used: invertible matrices are called *non-singular*.

Not every matrix has an inverse, as we have hinted. But when a matrix has an inverse, it has a unique inverse. To see this we need to think first about the size of an inverse. Suppose that A is $m \times n$. (We will see that $m = n$ when A has an inverse, but we won't use that at this point.) For $A \cdot C$ to be an identity matrix, the product $A \cdot C$ must be defined, and so C is $n \times$ (something). Since identity matrices are square, we see that $A \cdot C = I_m$, and so C is $n \times m$. If also $C \cdot A$ is an identity matrix, then we must have $C \cdot A = I_n$.

Now suppose that both C, D are inverses for the matrix A, and we will show that $C = D$. Indeed,

$$C = C \cdot I_m = C \cdot (A \cdot D) = (C \cdot A) \cdot D = I_n \cdot D = D$$

Because inverses are unique (when they exist!), we can denote the inverse of A with a notation that depends on the symbol A: we write A^{-1} for the inverse of A. We caution you that A^{-1} should not be read as $1/A$. We will stick to the definition: If A is $m \times n$ and has an inverse A^{-1}, then A^{-1} is $n \times m$ and

$$A^{-1} \cdot A = I_n \quad \text{and} \quad A \cdot A^{-1} = I_m$$

Example 1. Show that the following matrices are inverses.

$$A = \begin{pmatrix} 3 & 4 \\ 2 & 3 \end{pmatrix} \quad \text{and} \quad A^{-1} = \begin{pmatrix} 3 & -4 \\ -2 & 3 \end{pmatrix}$$

Example 2. Show that the following matrix A *is not invertible*, since $A \cdot C$ *cannot be* I_2, no matter what C is. (Hint: just look at the first column of AC and show you can't get $\begin{pmatrix} 1 \\ 0 \end{pmatrix}$.)

$$A = \begin{pmatrix} 1 & 2 \\ 2 & 4 \end{pmatrix}$$

Let's show that having an inverse allows us to solve $AX = B$ the way we do when A is a single number. The following proposition says that *if A has an inverse* (and not all matrices do!), *then* every right side to $AX = B$ has a solution, and that solution is unique.

PROPOSITION 5.1. *If the $m \times n$ matrix A has an inverse, then, for every $m \times 1$ matrix B, the equation $A \cdot X = B$ has a unique solution $X = A^{-1} \cdot B$.*

PROOF. Let A be invertible, so that A^{-1} is $n \times m$.

Given the $m \times 1$ matrix B, the matrix $A^{-1} \cdot B$ is defined and $n \times 1$. Compute

$$A \cdot (A^{-1} \cdot B) = (A \cdot A^{-1}) \cdot B = I_m \cdot B = B$$

and this shows that $A^{-1} \cdot B$ is a solution to $AX = B$. As for uniqueness, if $A \cdot X = B$, then

$$X = I_n \cdot X = (A^{-1} \cdot A) \cdot X = A^{-1} \cdot (A \cdot X) = A^{-1} \cdot B$$

This shows that X must be $A^{-1} \cdot B$.   □

You probably need to let the conclusion of Proposition 5.1 sink in. It imagines holding on to A and writing down various right sides B, solving various systems of equations that use A as the same coefficient matrix over and over. The assertion is that no matter which B is used, there is exactly one solution – of course, the solution varies with B.

In the previous proof, notice that both $A \cdot A^{-1} = I_m$ and $A^{-1} \cdot A = I_n$ were used. Inverses have to give the identity matrix on both sides.

What does Elimination say about $AX = B$ always having a unique solution? Here is where we discover that invertible matrices must be square.

1. Existence of the Inverse

PROPOSITION 5.2. *Let A be $m \times n$ and invertible. Then $m = n$, so that A is square, and no matter how the Elimination algorithm[1] is applied to A, its rank is n.*

PROOF. Because A has m rows, its rank in Elimination is less than or equal to m. Suppose that in some choice for Elimination, the rank of A comes out less than m. We will derive a contradiction to the fact that A has an inverse.

Our specific Elimination produces the row-echelon matrix A', and since the rank is less than m, the bottom row of the matrix A' is all 0's. Let D be a right side that is 0 except that it has a 1 at the bottom ($D[m, 1] = 1$ to be precise). Then the system $A' \cdot X = D$ is inconsistent, since its m-th equation says $0 = 1$.

Here is why that's interesting: To get from A to A', Elimination performed elementary operations, which, as we have seen, are reversible. In other words, there are elementary operations that would take us from A' back to A. Apply those elementary operations to the augmented matrix $[A'|D]$, and we end up with an augmented matrix of the form $[A|B]$. We know that the systems $A' \cdot X = D$ and $A \cdot X = B$ have exactly the same solutions. The former system has no solutions at all, and so the latter system $AX = B$ is inconsistent.

Recap of logic: if some choice of Elimination gives A rank less than m, then there is a right side B such that $A \cdot X = B$ is inconsistent. We are assuming that A is invertible; by Proposition 5.1, every right side gives a consistent system of equations. Therefore, there can be no such inconsistent B, and so A has rank m in every Elimination.

---

[1]Recall that there may be choices of pivot in Elimination. We are saying that no matter how those choices are made, the *number of pivots* will come out the same.

Next we show that A also has rank n. The system of equations $A \cdot X = \mathbb{O}_{m \times 1}$ has a unique solution by Proposition 5.1, once again using that A is invertible. For a system with n unknowns to have a unique solution, there cannot be any free unknowns in Elimination. Thus, the rank of A must be the number n of unknowns, no matter what choices are made in Elimination.

We have proved that the rank of A is both m and n, in every possible Elimination. In particular, $m = n$.   □

We are almost done. Proposition 5.2 *assumes* that A has an inverse and derives a conclusion. How do we know that A has or does not have an inverse? Fortunately, the converse of Proposition 5.2 is true. The proof of that fact shows how to use Elimination to find the inverse when it exists.

PROPOSITION 5.3. *Let A be an $n \times n$ matrix of rank n. Then A is invertible.*

PROOF. We begin by proving that if A is $n \times n$ and has rank n, then there is an $n \times n$ matrix C such that $A \cdot C = I_n$. The matrix C will turn out to be the inverse of A, and the next few paragraphs show how to compute this inverse.

Imagine a specific Elimination that gives A rank n. Then the system $A \cdot X = B$ is consistent for every possible right side B, since our chosen Elimination will never produce a row of 0's in A, and so $AX = B$ cannot be inconsistent. As in the proof of Proposition 5.2, this shows that the rank of A is n in every possible Elimination. Thus, we can drop the specific Elimination and imagine using the algorithm anyway we like.

As we did in Chapter 4, we will use as right sides the several columns of the identity matrix I_n. Make one big augmented matrix

$$[\, A \mid I_n \,]$$

Do Elimination, using A as coefficient matrix, and since A has rank n, its row-echelon form is I_n. Our augmented matrix turns into something that looks like this:

$$[\,I_n\,|\,C\,] \quad \text{where} \quad C \quad \text{is} \quad n \times n$$

As noted in Chapter 4 Elimination does not change the set of solutions: $A \cdot X = I_n$ is the same as $I_n \cdot X = C$. We see that $X = C$ is a solution to the second equation, and so it is a solution to the first equation: $A \cdot C = I_n$.

Here is what we've shown: if A is $n \times n$ of rank n, then there is an $n \times n$ matrix C such that $A \cdot C = I_n$. (And we've said that it will turn out that C is the inverse of A – we're almost there.)

Next we prove that C has rank n. This goes back to part of the argument in Proposition 5.2: consider the system $CX = \mathbb{O}_{n \times 1}$. This system is consistent, since $X = \mathbb{O}_{n \times 1}$ is a solution. We claim that this is the only solution to this equation. Indeed, if $CX = \mathbb{O}_{n \times 1}$, then compute

$$X = I_n \cdot X = (A \cdot C) \cdot X = A \cdot (C \cdot X) = A \cdot \mathbb{O}_{n \times 1} = \mathbb{O}_{n \times 1}$$

We see that $X = \mathbb{O}_{n \times 1}$ is the only possible solution. Since $CX = \mathbb{O}_{n \times 1}$ has a unique solution, the rank of C is n (there can be no free variables), as claimed.

Now we apply to C what we proved for A: since C has rank n, there is an $n \times n$ matrix D such that $C \cdot D = I_n$. In other words, the matrix C has an inverse. Because the inverse is unique, we have $A = D = C^{-1}$. In particular, $C = A^{-1}$, and we have shown that A has an inverse. $\square$

Let's try this out on an example:

$$A = \begin{pmatrix} 1 & 2 & 3 \\ 4 & 5 & 6 \\ 0 & 1 & 3 \end{pmatrix}$$

We augment our matrix A by I_3:

$$\begin{pmatrix} 1 & 2 & 3 & 1 & 0 & 0 \\ 4 & 5 & 6 & 0 & 1 & 0 \\ 0 & 1 & 3 & 0 & 0 & 1 \end{pmatrix}$$

Now we do Elimination, looking for pivots only in the first three columns (in the coefficient part).

$$\begin{pmatrix} 1 & 2 & 3 & 1 & 0 & 0 \\ 4 & 5 & 6 & 0 & 1 & 0 \\ 0 & 1 & 3 & 0 & 0 & 1 \end{pmatrix} -4 \cdot \text{I} + \text{II} \begin{pmatrix} 1 & 2 & 3 & 1 & 0 & 0 \\ 0 & -3 & -6 & -4 & 1 & 0 \\ 0 & 1 & 3 & 0 & 0 & 1 \end{pmatrix}$$

$$\begin{matrix} \text{II} \leftrightarrow \text{III} \\ -2 \cdot \text{II} + \text{I} \\ 3 \cdot \text{II} + \text{III} \end{matrix} \begin{pmatrix} 1 & 0 & -3 & 1 & 0 & -2 \\ 0 & 1 & 3 & 0 & 0 & 1 \\ 0 & 0 & 3 & -4 & 1 & 3 \end{pmatrix}$$

$$\begin{matrix} (1/3) \cdot \text{III} \\ 3 \cdot \text{III} + \text{I} \\ -3 \cdot \text{III} + \text{II} \end{matrix} \begin{pmatrix} 1 & 0 & 0 & -3 & 1 & 1 \\ 0 & 1 & 0 & 4 & -1 & -2 \\ 0 & 0 & 1 & -4/3 & 1/3 & 1 \end{pmatrix}$$

The row-echelon form of the coefficient matrix shows 3 pivots, and so the rank of A is 3. As in the proof of Proposition 5.3, the inverse of A appears in the rightmost 3 columns of the final augmented form:

$$\begin{pmatrix} 1 & 2 & 3 \\ 4 & 5 & 6 \\ 0 & 1 & 3 \end{pmatrix}^{-1} = \begin{pmatrix} -3 & 1 & 1 \\ 4 & -1 & -2 \\ -4/3 & 1/3 & 1 \end{pmatrix}$$

You can check that the products $A \cdot A^{-1}$ and $A^{-1} \cdot A$ are both equal to I_3.

To summarize: given an $n \times n$ matrix A, to see whether A has an inverse, do Elimination on the augmented matrix $[A|I_n]$ (with A as coefficient matrix). If the rank of A is less than n, then A does not have an inverse. If the rank of A is n, then row-echelon form will look like $[I_n|A^{-1}]$, showing the inverse in the rightmost n columns.

2. An Application: Polynomial Curve Fitting

All through applied work, we try to fit observed data to theoretical equations. In *regression*, for instance, we fit data points in the plane to a line. We want to think about the following general problem: we have m points (x_j, y_j), for $j = 1, 2, \ldots, m$ and trying to fit them to a polynomial of degree at most n;

let

$$(5.1) \qquad f(x) = c_0 + c_1 \cdot x + c_2 \cdot x^2 + \cdots + c_n \cdot x^n$$

In the problems at the end of Chapter 4 we already introduced this notation for a polynomial, where the c_j are turned into an $(n+1) \times 1$ matrix C.

Let V be the $m \times (n+1)$ Vandermonde matrix for $x_1, \ldots, x_m$ (see the problems at the end of Chapter 4). We know that

$$(5.2) \qquad V \cdot C = \begin{pmatrix} f(x_1) \\ f(x_2) \\ \vdots \\ f(x_m) \end{pmatrix}$$

and so to have the points (x_j, y_j) on the polynomial curve, we need $f(x_j) = y_j$ for each j. In other words, we need

$$V \cdot C = \begin{pmatrix} y_1 \\ y_2 \\ \vdots \\ y_m \end{pmatrix}$$

Write Y for the $m \times 1$ matrix on the right, and we need to have

$$(5.3) \qquad V \cdot C = Y$$

You know that two distinct points in the plane determine a unique line; the line will have the form $y = c_0 + c_1 \cdot x$ when the points have distinct x-coordinates.[2] Here is a generalization of this fact. Its proof uses our knowledge of matrix inverses.

PROPOSITION 5.4. *Suppose that (x_j, y_j), for $j = 1, 2, \ldots, m$ have distinct x-coordinates, and suppose that $n = m - 1$. Then there is a unique polynomial curve of degree at most n that passes through all these points.*

---

[2]For if the points have the same x-coordinate, then the line is vertical: of the form $x = a$, and it has no numerical slope.

PROOF. Since $n = m - 1$, the matrix V is square. A problem on p.67 in Chapter 4 shows that the equation $V \cdot X = \mathbb{O}$ has a unique solution. It follows that V has rank m, and so Proposition 5.3 shows us that it has an inverse. Then equation (5.3) has a unique solution C, and this solution defines the unique polynomial $f(x)$ of degree at most n, such that $f(x_j) = y_j$ for each j. □

In the case $m < n + 1$ (where we have relatively few points), it is not hard to see that there are infinitely many polynomials of degree at most n that fit the data points. We will skip over this in favor of the case that occurs most often in practice: $m > n + 1$ (relatively many points for relatively small degree). We have mentioned the regression line – that's the case when $n = 1$ and $m \geq 3$. Typically in this case the equation $V \cdot C = Y$ is inconsistent, so that there are no coefficients that satisfy the equations.[3] In this case the coefficients need to be *approximated* as best we can. The most common way to do this is to seek what is called the *best fit* polynomial of degree at most n.

To define what *best fit* means, assume we choose an $(n + 1) \times 1$ matrix C of coefficients arbitrarily. As above, the vector C corresponds to a polynomial $f(x)$. To indicate the dependence of f on C, we write $f(x) = f(C, x)$. Given points (x_j, y_j) for $1 \leq j \leq m$, we define the *squares error*

$$(5.4) \qquad E = \sum_{k=1}^{m} \big(f(C, x_k) - y_k\big)^2$$

There is a standard way to understand this: for each x-coordinate x_k we think of $f(C, x_k)$ as predicting where y_k *should be* – should be according to the coefficients c_j. The difference $f(C, x_k) - y_k$ is the difference between a *predicted* value $f(C, x_k)$ and the *observed* value y_k. The square of this $(f(C, x_k) - y_k)^2$

---

[3]It is important to note that there can be at least two reasons for this, neither mathematical. The data could be approximate, or the theory that predicts the coefficients could involve oversimplified assumptions. Either way, we would expect $f(x_j) \approx y_j$ and not be surprised if the two quantities are not equal.

gives a positive measure of this difference.[4] And the sum E adds up all these measures. Notice that E has the form $|u-v|^2$, where u represents the *predicted* values $f(C, x_k)$ and v lists the *observed* values y_k.

If $E = 0$, then since E is a sum of squares, we would have $f(C, x_k) = y_k$ for each k. In other words, if $E = 0$, then the polynomial curve fits the data exactly! Proposition 5.4 shows how to find C so that this exact fit occurs in the case that $m = n + 1$. In the present case $m > n + 1$, it makes sense to seek the *minimum* value of E – regarding C as variable. If the x-coordinates are distinct, it turns out that there is a unique matrix C of coefficients that minimizes E; the polynomial $f(C, x)$ is the *polynomial of degree at most n of best fit* to the given points (x_j, y_j). This fact is extremely important in applied work. We will state the formula for this polynomial and give a proof that the formula makes sense. It is not too hard to go farther and prove that the formula gives the polynomial of best fit; it is a messy calculation that we will defer in the interest of time. Once again we remind you: in the case $n = 1$, the *line* of best fit is the *regression line*.

The matrix $(V^T \cdot V)^{-1} \cdot V^T$ that occurs in the following is called the *Moore-Penrose inverse* of V. If V is invertible, then the Moore-Penrose inverse is just V^{-1}, but Moore-Penrose is used most often when V is not square. For the Moore-Penrose inverse to exist, the rank of V needs to be its number of columns.

---

[4]There are statistical reasons for using the squares; they involve advanced concepts that take a while to explain. This is unfortunate since the squares error is used so often. The best we can do is to reference Chapter 7 of [**14**].

PROPOSITION 5.5. *Given points (x_k, y_k) in the plane[5] for $1 \le k \le m$, with distinct x-coordinates, and given a positive integer n with $m \ge n+1$, there is a unique $(n+1) \times 1$ matrix C such that $f(C, x)$ is the polynomial of best fit to the points. In fact, if V is the $m \times (n+1)$ Vandermonde matrix for the x_k, then*

(5.5) $$C = (V^T \cdot V)^{-1} \cdot V^T \cdot Y$$

where $Y[i] = y_i$ for each i.

PROOF. We will prove only that $V^T \cdot V$ inverts, so that the formula for C makes sense. We will leave it to you to do the bookkeeping necessary to see that C is $(n+1) \times 1$. For instance, note that $V^T \cdot V$ is $(n+1) \times (n+1)$, and so it is square. To see that it inverts, consider the linear equation $V^T \cdot V \cdot X = \mathbb{O}$, where X is $(n+1) \times 1$. Multiply on the left by X^T to get $X^T \cdot V^T \cdot V \cdot X = 0$. Using properties of the transpose, this is

$$(V \cdot X)^T \cdot (V \cdot X) = 0 \quad \text{which is} \quad |V \cdot X|^2 = 0$$

It follows that $V \cdot X = \mathbb{O}$. Then, as noted in a problem in Chapter 4, this implies that $X = \mathbb{O}$.

We have shown that if $V^T \cdot V \cdot X = \mathbb{O}$, then $X = \mathbb{O}$. Since $V^T \cdot V$ is square, Proposition 5.3 then proves that it is invertible. $\square$

We remind you that Proposition 5.5 assumes that the x-coordinates of the data points are distinct. There is a version of the proposition in the case where some of the coordinates repeat, and there is a much more general version of the proposition that applies to more general, non-polynomial functions fit to data. This material is covered in books on advanced linear algebra and on statistics.

---

[5]Yet another technical detail: our proof needs real numbers x_j, y_j. There is a version of this proposition for complex numbers; we'll stick with the reals for now.

3. Problems

5.1.   For each of the following matrices, determine whether it has an inverse and, if it does, find that inverse.

a) $\begin{pmatrix} -2 & 3 & 5 \\ 1 & 2 & 1 \\ 1 & 1 & 0 \end{pmatrix}$
          b) $\begin{pmatrix} -1 & -1 & 1 \\ -1 & 1 & -1 \\ 1 & -1 & -1 \end{pmatrix}$
          c) $\begin{pmatrix} 1 & a & b \\ 0 & 1 & c \\ 0 & 0 & 1 \end{pmatrix}$

5.2.   Show that $\begin{pmatrix} a & b \\ c & d \end{pmatrix}$ has rank 2 if and only if $ad - bc \neq 0$. (Hint: Elimination! Consider two cases: first assume that $a \neq 0$, so that number can be used as a pivot. The other case: $a = 0$.)

5.3.   Let A and D be $n \times n$ and invertible. Show that $(AD)^{-1} = D^{-1}A^{-1}$.

5.4.   Let A be invertible and k a positive integer, then $(A^k)^{-1} = (A^{-1})^k$.

5.5.   Let A be invertible. Show that $(A^T)^{-1} = (A^{-1})^T$.

5.6.   Show that if A is 5×8, then there cannot be a matrix C such that $CA = I_8$. (Hint: if C does exist, how many solutions are there to $AX = \mathbb{O}_{5 \times 1}$?)

5.7.   Under what circumstances does the following matrix have an inverse?

$$\begin{bmatrix} a & b & c \\ 0 & d & e \\ 0 & 0 & f \end{bmatrix}$$

5.8.   Suppose that A is 7×5 and B is 7×1. Could $AX = B$ have a unique solution? Why or why not?

5.9.   Suppose that A is 3×6. Could $AX = B$ have a solution for all possible 3×1 matrices B? Why or why not?

5.10.   Find a 3×4 matrix A and 4×3 matrix C such that $A \cdot C = I_3$. (Note: it is not possible to have $C \cdot A = I_4$.)

5.11. Consider these data points: $(0, 2), (1, 3), (2, 5), (3, 4)$.

(a) Find the line $y = c_0 + c_1 \cdot x$ of best fit to the points, with its minimal squares error E_1.

(b) Find the parabola $y = c_0 + c_1 \cdot x + c_2 \cdot x^2$ of best fit, with its error E_2. (The numbers c_0, c_1 will be different for the parabola.) Why is it expected that $E_2 < E_1$?

(Note: the calculations are not horrendous by hand, but you might consider using software. Be sure to use Proposition 5.5, in any case.)

5.12. Suppose we have the equation $g \cdot x = y$, where g is a theoretical constant. Assume we have m data points (x_k, y_k) for $1 \le k \le m$ that have been observed. Find the minimum of the squares error $E(g)$ in this context.[6]

---

[6]Notes. The minimum is *not the numerical average of* y_k/x_k. Also, the expression $g \cdot x$ is not an arbitrary polynomial of degree at most 1, since there is no constant term, so this is not the kind of regression problem we discussed previously.

CHAPTER 6

The Determinant

The *determinant* is a number calculated in a certain definite way from the entries of a square matrix. According to the historian of mathematics D.E. Smith [**13**, p.476], Vandermonde, working in the late 1700's, was one of the first to come up with a general definition of the determinant and to begin to see its significance. By the early 1900's a standard book [**11**] compiling known facts about and uses of the determinant ran to 784 pages! It is hard to know how much to say about the determinant; it has many, many interesting properties, and it can be used extensively to do calculations. For instance, some people learn *Cramer's Rule* for solving linear equations – this computational algorithm uses various determinants to solve a system $A \cdot X = B$ in the case that A is square. Cramer also gives a computational formula for the inverse of an invertible matrix.[1] Another use of the determinant is in the calculation of *eigenvalues*, as we will see in Chapter 11.

In the interest of time we will take an operational approach to the determinant, showing how to compute it and stating a modest number of its more important properties. There are several ways to develop these properties rigorously, and you may well see them again at the advanced level.

To given the basic properties of the determinant, we need a definition: an $n \times n$ matrix A is *upper triangular* if the entries below the diagonal are 0.

---

[1]Our Elimination algorithm is more general than Cramer's formulas, requires less overhead to explain, and leads to better calculations when numerical approximations are used, and so we will stick to Elimination.

That means that if $1 \leq i < j \leq n$, then $A[i,j] = 0$.

We will assume, for each $n \times n$ matrix A, that there is a number $\det(A)$ such that

(1) The number $\det(A)$ is not 0 if and only if A is invertible.

(2) if A is upper triangular, then $\det(A)$ is the product of the entries on its diagonal; in particular, $\det(I_n) = 1$

(3) If A' is obtained from A by switching rows, then $\det(A) = -\det(A')$.

(4) If A' is obtained from A by adding a multiple of one row to another, then $\det(A) = \det(A')$.

(5) If A' is obtained from A by multiplying a row by the number $\alpha \neq 0$, then $\det(A) = (1/\alpha) \cdot \det(A')$.

(6) If A, B are $n \times n$, then $\det(A \cdot B) = \det(A) \cdot \det(B)$.

(7) $\det(A) = \det(A^T)$

Property (6) has a number of very remarkable consequences. If A is invertible, then $A \cdot A^{-1} = I_n$, so that property (6) says that

$$1 = \det(I_n) = \det(A \cdot A^{-1}) = \det(A) \cdot \det(A^{-1})$$

and we see that $\det(A^{-1}) = 1/\det(A)$. For instance, we see from this that $\det(A) \neq 0$ if A is invertible.

1. Computing the Determinant by Elementary Operations

The properties of the determinant listed above show directly how to use elementary operations to calculate the determinant. Let's show this by example. For each step, we list the elementary operation used and, in parenthesis, the property number used to justify that step.

$$\det \begin{pmatrix} 2 & 1 & 6 \\ 3 & 1 & 2 \\ 0 & 1 & 1 \end{pmatrix} = 2 \cdot \det \begin{pmatrix} 1 & 1/2 & 3 \\ 3 & 1 & 2 \\ 0 & 1 & 1 \end{pmatrix} \qquad 1/2 \cdot \text{I} \ (5)$$

$$= 2 \cdot \det \begin{pmatrix} 1 & 1/2 & 3 \\ 0 & -1/2 & -7 \\ 0 & 1 & 1 \end{pmatrix} \qquad -3 \cdot \text{I} + \text{II} \ (4)$$

$$= -2 \cdot \det \begin{pmatrix} 1 & 1/2 & 3 \\ 0 & 1 & 1 \\ 0 & -1/2 & -7 \end{pmatrix} \qquad \text{II} \leftrightarrow \text{III} \ (3)$$

$$= -2 \cdot \det \begin{pmatrix} 1 & 0 & 5/2 \\ 0 & 1 & 1 \\ 0 & 0 & -13/2 \end{pmatrix} \qquad \begin{matrix} -1/2 \cdot \text{II} + \text{I} \\ 1/2 \cdot \text{II} + \text{III} \end{matrix} \ (4)$$

$$= -2 \cdot 1 \cdot 1 \cdot (-13/2) = 13 \qquad (2)$$

The use of elementary operations recommends itself especially if we are asked to calculate the determinant of a matrix *numerically*. If you are interested in programming, you might try writing a determinant function using elementary operations.

2. Computing the Determinant with Cofactors

When we use the determinant in Chapter 11, we will have unknown entries in the matrices. The use of elementary operations will be hindered by this, and so we turn to a second method to compute determinants. This method of *cofactor expansion* is the one that most people learn in a course such as this one. The familiar formulas for the determinants of matrices that are 2×2 or 3×3 come from this approach. For example,

$$\det \begin{pmatrix} a & b \\ c & d \end{pmatrix} = ad - bc$$

There is an even simpler case that is usually not of interest: the determinant of a 1×1 matrix A is its entry: $\det(A) = A[1,1]$. The algorithms usually used for small size matrices are special, but they do generalize, as we now exhibit.

Calculation of the determinant by cofactor expansion is done inductively (recursively). We assume we know how to calculate the determinant of an $(n-1) \times (n-1)$ matrix, and we describe how to calculate the determinant of an $n \times n$ matrix. Since we know how to calculate the determinant of a 2×2 matrix, we can proceed from there to larger matrices.

In general, let A be $n \times n$ and assume we already know how to compute the determinant of $(n-1) \times (n-1)$ matrices. For each i, j with $1 \leq i \leq n$ and $1 \leq j \leq n$, define the i, j-*minor of A*, denoted $\mathbb{M}(i,j)$, to be the $(n-1) \times (n-1)$ matrix obtained by deleting the i-th row and j-th column from A. For example, if

$$(6.1) \qquad\qquad A = \begin{pmatrix} 1 & 2 & 3 \\ 4 & 5 & 6 \\ 7 & 8 & 9 \end{pmatrix}$$

then

$$\mathbb{M}(1,2) = \begin{pmatrix} 4 & 6 \\ 7 & 9 \end{pmatrix} \quad \text{and} \quad \mathbb{M}(3,3) = \begin{pmatrix} 1 & 2 \\ 4 & 5 \end{pmatrix}$$

The notation $\mathbb{M}(i,j)$ only makes sense if the matrix A has been specified. For brevity, we don't include the A in the notation for the minor.

The determinants of minors can be used to compute the determinant of A via the following formulas. Given a row i, we have

$$(6.2) \qquad\qquad \det(A) = \sum_{k=1}^{n} A[i,k](-1)^{i+k} \det(\mathbb{M}(i,k))$$

The signs $(-1)^{i+k}$ comes from the row and column of the minor considered. The terms in this sum are called *cofactors* and this is where the name, cofactor expansion, comes from.

Similarly, we can use a given column j to calculate the determinant.

$$(6.3) \qquad \det(A) = \sum_{k=1}^{n} A[k, j](-1)^{k+j} \det(\mathbb{M}(k, j))$$

Taking our example matrix (6.1), and choosing row 2 (rather arbitrarily), we compute

$$\det \begin{pmatrix} 1 & 2 & 3 \\ 4 & 5 & 6 \\ 7 & 8 & 9 \end{pmatrix} = 4(-1)^{2+1} \det(\mathbb{M}(2, 1))$$

$$+ 5(-1)^{2+2} \det(\mathbb{M}(2, 2)) + 6(-1)^{2+3} \det(\mathbb{M}(2, 3))$$

$$= 4(-1) \det \begin{pmatrix} 2 & 3 \\ 8 & 9 \end{pmatrix} + 5(1) \det \begin{pmatrix} 1 & 3 \\ 7 & 9 \end{pmatrix} + 6(-1) \det \begin{pmatrix} 1 & 2 \\ 7 & 8 \end{pmatrix}$$

$$= -4(18 - 24) + 5(9 - 21) - 6(8 - 14) = 0$$

The same determinant can be computed choosing, say, column 3.

$$\det \begin{pmatrix} 1 & 2 & 3 \\ 4 & 5 & 6 \\ 7 & 8 & 9 \end{pmatrix} = 3(-1)^{1+3} \det(\mathbb{M}(1, 3))$$

$$+ 6(-1)^{2+3} \det(\mathbb{M}(2, 3)) + 9(-1)^{3+3} \det(\mathbb{M}(3, 3))$$

$$= 3 \det \begin{pmatrix} 4 & 5 \\ 7 & 8 \end{pmatrix} - 6 \det \begin{pmatrix} 1 & 2 \\ 7 & 8 \end{pmatrix} + 9 \det \begin{pmatrix} 1 & 2 \\ 4 & 5 \end{pmatrix}$$

$$= 3(-3) - 6(-6) + 9(-3) = 0$$

A very important fact: you get the same determinant no matter which row or column is chosen for cofactor expansion. In other words, all the formulas, (6.2) and (6.3), give the same number every time for the same matrix.[2] This fact can be listed along with the properties (1)-(7) above, and you then have what are considered the basic facts about the determinant. In our course, we will have time only to sketch the logical train that establishes these facts, and only to prove a couple of implications along that train. This is a situation in

---

[2]You can show that this is the case for 2×2 matrices; there are four possible cofactor expansions.

which the mathematician in your instructor has been overruled by the term schedule – a rare disagreement.

So, here's a sketch of how the facts may be established: First, it is easy to show that the determinant of a 1×1 matrix satisfies the necessary properties. (Most of them don't even apply to that case – you can't switch rows for example.) For an induction argument, we assume, for some specific integer $n \geq 2$ that the determinant of $(n-1) \times (n-1)$ matrices is defined to have all the properties that apply. Taking as definition of the determinant the cofactor expansion about row 1, we show that properties (2), (3), (4), and (5) hold, so that we know the effect of elementary operations on the determinant. Property (1) follows from applying Elimination to see whether the given matrix is invertible. The use of elementary operations allows us to show that the other row cofactor expansions give the same determinant. The cofactor expansions then establish (7), and we thereby obtain the column cofactor expansions. Property (6) is a little harder to obtain – it comes from understanding the effect of elementary operations on the identity matrix.

3. Some Properties of the Determinant

Property (2) gives the determinant of each upper triangular matrix. There is a similar fact for *lower triangular matrices*. An $n \times n$ matrix A is lower triangular if it is zero above its diagonal. The transpose of A is upper triangular and has the same diagonal at A. By Property (7), we have $\det(A) = \det(A^T)$, and Property (2) says that $\det(A^T)$ is the product of the diagonal entries.

Here is how Property (1) follows from the information about elementary operations.

PROPOSITION 6.1. *Let A be an $n \times n$ matrix. Then $\det(A) \neq 0$ if and only if A is invertible.*

PROOF. Perform elimination on A to get a row echelon form B. Properties (3), (4), (5) show that there is a non-zero number α such that $\det(A) = \alpha \cdot \det(B)$.

If A is invertible, then $B = I_n$, so that $\det(A) = \alpha \neq 0$.

If A is not invertible, then B has a row of 0's. We will show that $\det(B) = 0$, and it will follow that $\det(A) = 0$. Suppose that row j of B is all 0's. Form the matrix C by multiplying row j of B by 2. Property (5) says that $\det(C) = 2 \cdot \det(B)$. On the other hand, since row j of B is all 0's, we have $C = B$, and so $\det(B) = 2 \cdot \det(B)$, so that $\det(B) = 0$ as needed. $\square$

This might be a good time to collect the several conditions equivalent to A being invertible. Given an $n \times n$ matrix A, the following are equivalent (they are all true or they are all false):

(1) A has an inverse.

(2) The rank of A is n (no matter how Elimination is done).

(3) The nullity of A is 0 (no matter how ...)

(4) $\det(A) \neq 0$.

(5) $AX = B$ is consistent for all $n \times 1$ matrices B.

(6) For each $n \times 1$ matrix B the equation $AX = B$ has a unique solution.

(7) The only solution to $AX = \mathbb{O}_{n \times 1}$ is $X = \mathbb{O}_{n \times 1}$.

4. Problems

6.1. For each of the following matrices, compute the determinant. Do at least one of them using Elimination, and do at least one using cofactors. (You might want to check your work using a calculator or computer, but do the calculations by hand to make sure you understand the formulas.)

a) $\begin{pmatrix} 2 & 5 & 0 & 0 \\ 1 & 2 & -3 & 1 \\ 3 & 8 & 2 & -1 \\ 1 & 3 & -10 & 3 \end{pmatrix}$
b) $\begin{pmatrix} 3 & 0 & -4 & 0 \\ 1 & 1 & 3 & -1 \\ 1 & 0 & 6 & -5 \\ 1 & 2 & 1 & 3 \end{pmatrix}$
c) $\begin{pmatrix} -2 & 1 & 3 \\ 1 & 3 & 1 \\ 3 & 0 & -2 \end{pmatrix}$

6.2. Let A be $n \times n$ and suppose that row 1 of A is multiplied by the number α to form the matrix A'. Show that $\alpha \cdot \det(A) = \det(A')$. (Hint: cofactors about row 1.)

6.3. Consider the general 3×3 matrix A. Compute the determinant using cofactors about two different rows and about one column (your choice), and show that you get the same answer in all three cases. (Of course, this answer should agree with the "crisscross" formula for the determinant of a 3×3 matrix.)

6.4. Suppose that A and B are $n \times n$ matrices and assume that AB is invertible. Show that A and B are each invertible. (Hint: $\det(A \cdot B)$?)

6.5. Let A be $n \times n$ and let β be a number. Show that $\det(\beta \cdot A) = \beta^n \cdot \det(A)$. (Hint: $\beta \cdot A$ is obtained from A by a succession of row multiplications.)

6.6. Complete the following steps to show that if P is the parallelogram with corners at $(0,0)$ and (a,b) and (c,d) and $(a+c, b+d)$, then the area of P is the absolute value of the determinant of

$$A = \begin{pmatrix} a & c \\ b & d \end{pmatrix}$$

a) Let L be the ray from $(0,0)$ through (a,b), and let θ be the angle from this ray to the positive x-axis. Rotate P by the angle θ. Argue that one side of the rotated parallelogram is on the positive x-axis.

b) Recall the rotation matrix $R(\theta)$. Argue that the columns of

$$R(\theta) \cdot A$$

are corners of the rotated parallelogram, and so $R(\theta) \cdot A$ has the form

$$\begin{pmatrix} e & f \\ 0 & g \end{pmatrix}$$

c) Show that the area of the rotated parallelogram is $|e \cdot g|$.

d) Complete the following calculation to finish the problem.

$$|e \cdot g| = \left| \det \begin{pmatrix} e & f \\ 0 & g \end{pmatrix} \right| = |\det (R(\theta) \cdot A)| = \cdots$$

CHAPTER 7

Constant Coefficient DE's

Mathematics arises when several individual problems admit of the same abstract description. The three crude pictures here represent three examples that will be explained momentarily. In physical terms, each of these examples exhibits *forced, damped oscillations.*

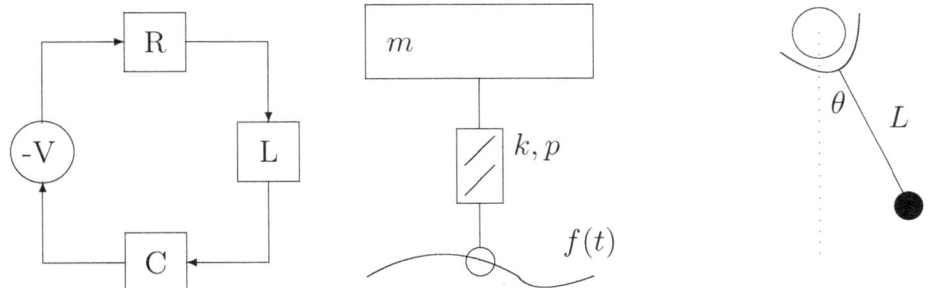

Figure 1: Three Examples.

Example 1: RLC Circuit. The leftmost diagram in Figure 1 depicts a simple linear electrical circuit as in Chapter 4 but with *components* depicted by the boxes. Each component has a constant associated with it, written in the box; we will explain these constants momentarily. The arrows give the positive direction for the flow of current. As we explained in Chapter 4, in each edge of the diagram there is a current and a potential drop, and there are Kirchoff equations for each of the rules. In the present situation, the current and potential are functions of time t.

We can describe each component by describing the potential drop that it affords: The component labeled $-V$ is a *potential source* (a battery, a signal, etc.), and its drop in potential is $-V$. (By convention $-V$ is used rather than V. Ask the physicists.) The box labeled R is a *resistor*, and its drop in potential is defined to be $R \cdot I$, where I is the current. The box L is an *inductor*, and its drop is $L \cdot I'$, where, again, I is current. The C box is a *capacitor*, and if p is its drop, then $C \cdot p' = I$. In our diagram, the current rule shows that the current I is the same in each edge. There is only one loop, and so we have one equation for the voltage rule:

$$-V + R \cdot I + L \cdot I' + p = 0$$

We'll put the V on the right side. If we take the derivative of this equation, we'll be able to use I in the capacitor term:

(7.1) $$R \cdot I' + L \cdot I'' + \frac{1}{C} \cdot I = V'$$

Dividing by L and rearranging in derivative order, we have

(7.2) $$I'' + \frac{R}{L} \cdot I' + \frac{1}{LC} \cdot I = \frac{1}{L} \cdot V'$$

We regard V as given and we want to solve for I. Because the highest derivative of I involved is the second derivative, this is called a *second-order* differential equation. The coefficients on I and I' are constant, and so this is a *constant coefficient* problem. Not surprisingly, the name *RLC circuit* comes from the names for the three constants used.

Example 2: A Shock Absorber. The diagram in the middle of Figure 1 represents a heavy object (such as a car) of mass m supported by a spring inside a *dashpot* – a cylinder filled with fluid. Think shock absorber. The curve at the bottom is the terrain over which the object moves. The circle at the base of the spring is a frictionless contact with the ground (perhaps it represents one of the tires). The variable y measures the height of the car above the x-axis – the x-axis acting as usual as a horizontal reference. The height of the ground above the x-axis is described by the curve $f(t)$.

We assume that the following accelerations are in play: a constant gravitational acceleration $-g$, the acceleration from the spring proportional to the displacement of the spring from its natural length, and a resistance from the dashpot fluid proportional to y'. For the spring, at a given point in time, the spring length is $y - f(t)$. If the natural length of the spring is L, then the displacement from natural length is $y - f(t) - L$, and so the acceleration has the form $-k \cdot (y - f(t) - L)$ for some positive constant[1] k. For the dashpot, the resistance has the form $-p \cdot y'$ for some positive constant p. Putting the accelerations together, we find that

$$y'' = -g - p \cdot y' - k \cdot (y - f(t) - L)$$

which is

(7.3) $$y'' + p \cdot y' + k \cdot y = -g + k \cdot (f(t) + L)$$

The mathematician notices that (7.3) has the same form as (7.1): second order DE with constant coefficients on the left.

Example 3: A Pendulum Clock. The diagram at the right of Figure 1 stands for the pendulum apparatus of a clock. The number L is the length of the pendulum arm. We will assume a constant gravitational acceleration $-g$ (down, of course) and that there is a small resistance to the pendulum motion proportional to the angular velocity θ'. It is an easy calculation of physics that the effect of gravity on θ is to induce acceleration $-(g/L) \cdot \sin(\theta)$. Typical pendulum clocks are designed to have $|\theta| < \pi/18$, and in this range $\sin(\theta) \approx \theta$ is a good approximation, and our gravitational acceleration is approximately $-(g/L) \cdot \theta$. The curved shape at the top of the pendulum arm is an *escapement* that is connected to the circle at the top – meant to represent a spring

---

[1]The constant k is the physical spring constant divided by the mass of the object.

that imparts an impulse to the pendulum when $\theta = \pm\theta_0$. We represent the acceleration from this impulse as $f(t)$, and, remembering the resistance, we approximate

$$\theta'' = -\frac{g}{L} \cdot \theta - k \cdot \theta' + f(t) \quad \text{which is} \quad \theta'' + k \cdot \theta' + \frac{g}{L} \cdot \theta = f(t)$$

The same as (7.1) and (7.3)! ∎

In each of these three examples, we obtain a DE in the following form:

$$(7.4) \qquad\qquad y'' + p \cdot y' + q \cdot y = g(t)$$

where p, q are non-negative constants. This general equation is an example of a *constant coefficient DE*. To write the general form of such an equation, we need to recall the notation $y^{(k)}$ for the k-th derivative of the function y. Here is the general *constant coefficient* DE:

$$(7.5) \qquad y^{(n)} + a_{n-1} \cdot y^{(n-1)} + \cdots + a_2 \cdot y'' + a_1 \cdot y + a_0 \cdot y = g(t)$$

where $a_0, a_1, \ldots, a_{n-1}$ are constants, and $g(t)$ is continuous. This formidable expression is an *n-th order* DE, since the highest derivative of y is the n-th. An initial value problem in this context gives n values at $t = 0$, starting with y_0:

$$y_0, \; y_0', \; y_0'', \; \ldots, \; y_0^{(n-1)}$$

It is easier to remember that an IVP gives n values than it is to remember that the $n - 1$-st derivative is the last one in the list.

Perhaps this is a good time to say something about the mathematician's tendency to ignore distinctions between objects in favor of mathematical similarities. One might say, "Electrical circuits are not shock absorbers, and shock absorbers are not clocks. Don't we lose something *essential* when we think of them as giving the same DE?" The mathematician replies, "If the differences between the objects in question are important, those differences should be reflected in the abstract model for each case. The models are the business of the discipline in which the objects are studied; the mathematician starts with whatever model is given. In the examples given, the model is a DE; the fact that the DE's were essentially the same is an *insight* about the three models. There is something about a circuit that is like a shock absorber – that's an insight that *results* from abstract consideration. It might well turn out that we need to include more information in one of the models to understand the behavior of that object particularly. A mathematician, as a mathematician, studies each abstract model for its own sake – if you want information that *distinguishes* the behavior of an RLC circuit from that of a clock, then give us more particular models!" You should think about these different points of view; each recommends itself in an interesting way.

There are several methods for solving constant coefficient problems – each method has its own advantages and disadvantages. The various methods involve the same basic idea, more or less: reduce the DE to an *algebraic* equation. In fact, the algebra involved in each method is more or less the same, as well. We will be studying a single approach that has an echo in linear algebra and that introduces an important idea: the *differential operator*. You are referred to the standard DE books (in the bibliography) for alternate approaches.

1. The Differential Operator

We introduce an idea that will lead to solutions of constant coefficient problems. This idea is simply to use the letter D to represent taking the derivative with respect to t, but we will see that D begins to behave algebraically. Thus,

$$D\cos(2t) = -2 \cdot \sin(2t), \quad Dt^2 = 2t, \quad \text{and so on}$$

The second derivative takes the derivative twice!

$$y'' = D(y') = D(D(y)) = D^2 y$$

Then $D^3 y$ represents the third derivative of y, and so on. This notation will allow us to form polynomials in D. Here's an example:

$$y^{(3)} - 4 \cdot y'' + 7 \cdot y = t^2$$
$$D^3 y - 4 \cdot D^2 \cdot y + 7 \cdot y = t^2$$
$$(D^3 - 4 \cdot D^2 + 7) \cdot y = t^2$$

Notice that the right side t^2 just sits there. We are using powers of D and coefficients to represent a combination of derivatives of y. We will see that *roots* of the polynomial in D hold the key to solutions of the DE.

How about a general example of order 2? Let b, c be constants.

$$y'' + b \cdot y' + c \cdot y = 0$$
$$D^2 y + b \cdot D y + c \cdot y = 0$$
$$(D^2 + b \cdot D + c) \cdot y = 0$$

Notice that the polynomial $D^2 + b \cdot D + c$ has degree 2, since its highest power of D is 2. That power of D comes from the second derivative of y, and that second derivative is the highest derivative. We are pointing out that the *degree* of the polynomial is the *order* of the DE.

The symbol D in this context is called an *operator* – that's the name for a function that operates on functions. The symbol D operates on functions by taking their derivative. The polynomials we obtained involving powers of D are called *operator polynomials*.

Here is what the DE (7.5) looks like using an operator polynomial.

$$(7.6) \qquad (D^n + a_{n-1} \cdot D^{n-1} + \cdots + a_2 \cdot D^2 + a_1 \cdot D + a_0) \cdot y = g(t)$$

Let's show how operator polynomials help us find solutions. Let c be a complex number constant and recall that the function e^{ct} is a solution to the equation $y' = c \cdot y$. Expressing this with D, this is $D \cdot y = c \cdot y$ and that's $(D - c) \cdot y = 0$. We're saying that e^{ct} is a solution to $(D - c) \cdot y = 0$. Notice that

c is a *root* of the polynomial $D - c$ and that e^{ct} is a *solution* to $(D - c) \cdot y = 0$. Here's how this fact can be used. Consider the DE

$$y'' + 2 \cdot y' - 8 \cdot y = 0$$

We write y and y' and y'' using D:

$$D^2 \cdot y + 2 \cdot D \cdot y - 8 \cdot y = 0$$
$$(D^2 + 2 \cdot D - 8) \cdot y = 0$$

We can factor the polynomial in D.

$$(D + 4) \cdot (D - 2) \cdot y = 0$$

Notice the roots -4 and 2 of the operator polynomial. Since $(D - 2)e^{2t} = 0$, the function e^{2t} is a solution to our second-order DE:

$$(D + 4) \cdot (D - 2) \cdot e^{2t} = (D + 4) \cdot 0 = 0$$

The other factor also gives us a solution: $(D + 4)e^{-4t} = 0$, so that

$$(D + 4) \cdot (D - 2) \cdot e^{-4t} = (D - 2) \cdot (D + 4) \cdot e^{-4t} = (D - 2) \cdot 0 = 0$$

In the next section, we'll leverage this simple idea. For now we repeat: for each number c, we have $(D - c) \cdot e^{ct} = 0$, and so when c is a root of the operator polynomial $f(D)$ (when $D - c$ is a factor of $f(D)$), the function e^{ct} is a solution to $f(D) \cdot y = 0$.

2. Homogeneous Problems

In the previous section we wrote the DE (7.5) using an operator polynomial (7.6). Here's what this looks like abstractly.

$$f(D) \cdot y = g(t)$$

where $f(D)$ is the operator polynomial. This equation is *homogeneous*[2] if $g(t) = 0$ (that's an identity – zero for all t). In homogeneous problems, the factoring idea of the previous section will give us all possible solutions. In section 4, we'll consider non-zero $g(t)$.

We need a couple of general facts about the general homogeneous constant coefficient problem: $f(D) \cdot y = 0$.

Principle of scaling. If u is a solution to $f(D) \cdot y = 0$, and if c is a constant, then $c \cdot u$ is also a solution.

This is because $D(c \cdot u) = c \cdot Du$, and so $D^2(cu) = c \cdot D^2 u$, and so on. Thus, $f(D)(c \cdot u) = c \cdot f(D)u$, and so if $f(D)u = 0$, then

$$f(D)(cu) = c \cdot f(D)u = c \cdot 0 = 0$$

Principle of superposition If u and v are solutions to $f(D) \cdot y = 0$, then so is $u + v$.

This goes back to the derivative being linear: $D(u+v) = u'+v' = Du+Dv$. It follows that $f(D)(u+v) = f(D) \cdot u + f(D) \cdot v$, and so if $f(D)u = 0 = f(D)v$, then $u + v$ gives 0 too.

Let's do a specific example and see the two principles in action. In the last section we showed that e^{2t} and e^{-4t} are solutions to $(D^2 + 2D - 8) \cdot y = 0$. Scaling then says that $A_1 \cdot e^{2t}$ and $A_2 \cdot e^{-4t}$ are solutions, for arbitrary constants A_1, A_2. Superposition then says that

(7.7) $y = A_1 \cdot e^{2t} + A_2 \cdot e^{-4t}$

is a solution for all A_1, A_2. That's a lot of solutions!

Recall that an IVP for $(D^2+2D-8) \cdot y = 0$ would specify two value: y_0, y_0'. (Two initial values since two is the degree of the operator polynomial and it is

---

[2]The word *homogeneous* is used in a variety of ways in mathematics. We are not sure about the origin of its use here; we suspect it is because of the principle of scaling mentioned in this section.

the order of the DE.) Suppose we want $y_0 = 3$ and $y_0' = -2$. Will our solution (7.7) give us the initial values? Observe that

$$y = A_1 \cdot e^{2t} + A_2 \cdot e^{-4t} \quad \text{so that} \quad 3 = y_0 = A_1 + A_2$$

$$y' = 2 \cdot A_1 \cdot e^{2t} - 4 \cdot A_2 \cdot e^{-4t} \quad \text{so that} \quad -2 = y_0' = 2 \cdot A_1 - 4 \cdot A_2$$

The initial conditions give us a system of linear equations:

$$(7.8) \qquad \begin{pmatrix} 1 & 1 \\ 2 & -4 \end{pmatrix} \cdot \begin{pmatrix} A_1 \\ A_2 \end{pmatrix} = \begin{pmatrix} 3 \\ -2 \end{pmatrix}$$

Elimination solves this $A_1 = 5/3$ and $A_2 = 4/3$, and we have a solution to the IVP:

$$y = \frac{5}{3} \cdot e^{2t} + \frac{4}{3} \cdot e^{-4t}$$

There is a larger point to make about the calculation of A_1, A_2. The coefficient matrix in the linear equation (7.8) has determinant $-6 \neq 0$, and so that matrix is invertible. This means the system in (7.8) is consistent *no matter what right side is used*. The right side came directly from the initial values. Thus, no matter what initial conditions we put on a solution to the DE here, we will be able to solve for (unique) values of A_1, A_2. This tells us that the general solution (7.7) solves all possible IVP's for the DE. The method we are teaching in this section will have that property in every homogeneous constant coefficient DE.

Let's do another example.

Problem. Solve the IVP

$$y^{(3)} + 4 \cdot y'' + 3 \cdot y' = 0 \quad \text{and} \quad y_0 = 6, \ y_0' = -9, \ y_0'' = 21$$

Also, show that the general solution will solve all possible IVP's.

Solution. The operator polynomial: $(D^3 + 4 \cdot D^2 + 3 \cdot D)y = 0$, and we factor

$$D^3 + 4 \cdot D^2 + 3 \cdot D = D \cdot (D^2 + 4 \cdot D + 3) = D \cdot (D+3) \cdot (D+1)$$

Thus, the roots are $0, -3, -1$, so we have solutions $e^{0t} = 1$ and e^{-3t} and e^{-t}. Scaling and superposition give a general solution

$$y = A_1 + A_2 \cdot e^{-3t} + A_3 \cdot e^{-t}$$

for arbitrary constants A_1, A_2, A_3. Now we include the initial conditions. To compute y_0' and y_0'', we need the derivative and second derivative of y:

$$y = A_1 + A_2 \cdot e^{-3t} + A_3 \cdot e^{-t} \quad \text{and} \quad 6 = y_0 = A_1 + A_2 + A_3$$
$$y' = -3 \cdot A_2 \cdot e^{-3t} - A_3 \cdot e^{-t} \quad \text{and} \quad -9 = y_0' = -3A_2 - A_3$$
$$y'' = 9 \cdot A_2 \cdot e^{-3t} + A_3 \cdot e^{-t} \quad \text{and} \quad 21 = y_0'' = 9 \cdot A_2 + A_3$$

The initial conditions give a system of linear equations (again!).

$$\begin{pmatrix} 1 & 1 & 1 \\ 0 & -3 & -1 \\ 0 & 9 & 1 \end{pmatrix} \cdot \begin{pmatrix} A_1 \\ A_2 \\ A_3 \end{pmatrix} = \begin{pmatrix} 6 \\ -9 \\ 21 \end{pmatrix}$$

We can solve this system: $A_1 = 1$, $A_2 = 2$, $A_3 = 3$, and our solution is

$$y = 1 + 2 \cdot e^{-3t} + 3 \cdot e^{-t}$$

Finally, the coefficient matrix in the system that gives the initial values has determinant $6 \neq 0$, and so we would be able to solve the system for every possible right side (for every possible IVP).

The factor $D - c$ gives solution e^{ct} even when c is not real.

Problem. Solve the IVP

$$y'' + 4 \cdot y = 0, \quad \text{and} \quad y_0 = 3, \ y_0' = 2$$

Solution. The DE is $(D^2 + 4) \cdot y = 0$. We can factor $D^2 + 4 = (D + 2i) \cdot (D - 2i)$, and the roots are $-2i$ and $2i$. We get a general solution

$$y = A_1 \cdot e^{2i \cdot t} + A_2 \cdot e^{-2i \cdot t}$$

Because we have non-real roots, we may have to use non-real coefficients A_1, A_2. This will come out naturally as we solve the IVP.

$$y = A_1 \cdot e^{2i \cdot t} + A_2 \cdot e^{-2i \cdot t} \quad \text{and} \quad 3 = y_0 = A_1 + A_2$$

$$y' = 2i \cdot A_1 \cdot e^{2i \cdot t} - 2i \cdot A_2 \cdot e^{-2i \cdot t} \quad \text{and} \quad 2 = y_0' = 2i \cdot A_1 - 2i \cdot A_2$$

The system of equations:

$$\begin{pmatrix} 1 & 1 \\ 2i & -2i \end{pmatrix} \cdot \begin{pmatrix} A_1 \\ A_2 \end{pmatrix} = \begin{pmatrix} 3 \\ 2 \end{pmatrix} \quad \text{solution:} \quad A_1 = \frac{3-i}{2}, \; A_2 = \frac{3+i}{2}$$

and

$$y = \frac{3-i}{2} \cdot e^{2i \cdot t} + \frac{3+i}{2} \cdot e^{-2i \cdot t}$$

Yet again, the determinant of the coefficient matrix used to get A_1, A_2 is non-zero, and so we can solve all IVP's.

The solution we obtained in the last problem is fine, but it would be very common to put the solution in cosine/sine form or in cosine form. We'll digress to remind you how to get cosine/sine form.

$$\begin{aligned} y &= \frac{3-i}{2} \cdot e^{2i \cdot t} + \frac{3+i}{2} \cdot e^{-2i \cdot t} \\ &= \frac{3-i}{2} \cdot (\cos(2t) + i \cdot \sin(2t)) + \frac{3+i}{2} \cdot (\cos(2t) - i \cdot \sin(2t)) \\ &= \left(\frac{3-i}{2} + \frac{3+i}{2} \right) \cdot \cos(2t) \\ &\quad + \left(\frac{3-i}{2} \cdot i - \frac{3+i}{2} \cdot i \right) \cdot \sin(2t) \\ &= 3 \cdot \cos(2t) + \sin(2t) \end{aligned}$$

This form of the solution: $y = 3 \cdot \cos(2t) + \sin(2t)$ hides the non-real numbers used to get the solution.

Now we come to a harder problem. Consider the following IVP.

$$y'' - 6 \cdot y' + 9 \cdot y = 0, \quad \text{and} \quad y_0 = 0, \; y_0' = 1$$

The operator polynomial $(D^2 - 6 \cdot D + 9) \cdot y = 0$, and this factors $(D-3)^2 y = 0$. The wrinkle here is that the root 3 is repeated. We get e^{3t} as solution, and we can write $y = A \cdot e^{3t}$ where A is a constant. The trouble is that we cannot solve the IVP, for $y_0 = 0$ implies that $A = 0$, and $y = 0$ does not satisfy $y_0' = 1$. In the next section we find our way out of this difficulty.

3. Repeated Roots

We begin with a calculation: let $p(t)$ be a polynomial, let c be a complex number, and compute the following, using the product rule.

$$D\big(p(t) \cdot e^{ct}\big) = p'(t) \cdot e^{ct} + p(t) \cdot c \cdot e^{ct}$$

We use this to perform another calculation:

$$
\begin{aligned}
(D - c)\big(p(t) \cdot e^{c \cdot t}\big) &= D\big(p(t) \cdot e^{c \cdot t}\big) - c \cdot p(t) \cdot e^{c \cdot t} \\
&= p'(t) \cdot e^{c \cdot t} + p(t) \cdot c \cdot e^{c \cdot t} - c \cdot p(t) \cdot e^{c \cdot t} \\
&= p'(t) \cdot e^{c \cdot t}
\end{aligned}
$$

Summary: $(D-c) \cdot \big(p(t) \cdot e^{c \cdot t}\big) = p'(t) \cdot e^{c \cdot t}$. Notice that the exponential function stays put; the polynomial undergoes the derivative. We can apply this identity twice:

$$
\begin{aligned}
(D - c)^2\big(p(t) \cdot e^{c \cdot t}\big) &= (D - c) \cdot (D - c)\big(p(t) \cdot e^{c \cdot t}\big) \\
&= (D - c) \cdot \big(p'(t) \cdot e^{c \cdot t}\big) \\
&= p''(t) \cdot e^{c \cdot t}
\end{aligned}
$$

We can keep this up to higher powers of $D - c$ as well. If we have k factors of $D - c$ we have, we end up with k derivatives of $p(t)$.

Recall the IVP that we were not able to do in the previous section: $(D - 3)^2 \cdot y = 0$ and $y_0 = 0$ and $y_0' = 1$. Let's try $y = p(t) \cdot e^{3t}$, where $p(t)$ is a polynomial. We have

$$(D - 3)^2\big(p(t) \cdot e^{3t}\big) = p''(t) \cdot e^{3t}$$

To get a solution to the DE $(D-3)^2 y = 0$, we need $p''(t) = 0$. Any function of the form $p(t) = A \cdot t + B$ satisfies $p''(t) = 0$, and so any function of the form

$$y = (A \cdot t + B) \cdot e^{3t}$$

is a solution to the equation $(D-3)^2 \cdot y = 0$. This is getting abstract, so you might want to perform the (tedious!) calculation of showing that $(A \cdot t + B) \cdot e^{3t}$ really is a solution to the DE we started with:

$$y'' - 6 \cdot y' + 9 \cdot y = 0$$

Continuing, let's see that this function can be molded to the initial conditions $y_0 = 0$, $y_0' = 1$.

$$0 = y_0 = (A \cdot 0 + B) \cdot e^{3 \cdot 0} = B$$

This simplifies $y = A \cdot t \cdot e^{3t}$. To apply the other initial value $y_0' = 1$, we calculate

$$y' = A \cdot e^{3t} + A \cdot t \cdot 3 \cdot e^{3t}$$

and so

$$1 = y_0' = A$$

Thus, $y = t \cdot e^{3t}$ solves the IVP.

We could carry the abstract derivation further, but we'll skip to describing what factor goes with what general solution degree by degree, ending with the general pattern.

$D - c$	goes with	$A_1 \cdot \exp(c \cdot t)$
$(D - c)^2$	goes with	$(A_1 + A_2 \cdot t) \cdot \exp(c \cdot t)$
$(D - c)^3$	goes with	$(A_1 + A_2 \cdot t + A_3 \cdot t^2) \cdot \exp(c \cdot t)$
$\vdots$	$\vdots$	$\vdots$
$(D - c)^k$	goes with	$(A_1 + A_2 \cdot t + \cdots + A_k \cdot t^{k-1}) \cdot \exp(c \cdot t)$

When a factor occurs k times, there are k constants associated with it, one constant for each power of t, starting with the 0-th power. Let's include this idea in another example.

Problem. Solve the IVP

$$y^{(3)} - 3 \cdot y' - 2 \cdot y = 0 \quad \text{and} \quad y_0 = 6, \ y'_0 = -1, \ y''_0 = 14$$

Solution. The DE is $(D^3 - 3 \cdot D - 2)y = 0$. To factor this, we need to get lucky. We stumble on the fact that 2 is a root, and so we get $(D-2)(D^2+2D+1)y = 0$, and this is $(D-2)(D+1)^2 y = 0$. We might talk through the counting of coefficients that goes with each factor: $D-2$ occurs *once* and so its part of the solution has *one* coefficient: $A_1 \cdot e^{2t}$. The factor $D+1$ occurs *twice*, and so its part in the solution has *two* coefficients: $(A_2 + A_3 \cdot t) \cdot e^{-t}$. Putting these together, the general solution appears:

$$y = A_1 \cdot e^{2t} + \left(A_2 + A_3 \cdot t\right) \cdot e^{-t}$$

We pause to note: DE of order 3, operator polynomial of degree 3, general solution with 3 arbitrary constants. Check. The initial conditions need y' and y''.

$$y = A_1 \cdot e^{2t} + \left(A_2 + A_3 \cdot t\right) \cdot e^{-t} \quad \text{and} \quad 6 = y_0 = A_1 + A_2$$
$$y' = 2 \cdot A_1 \cdot e^{2t} + \left(A_3 - A_2 - A_3 \cdot t\right) \cdot e^{-t}$$
$$\text{and} \quad -1 = y'_0 = 2A_1 + A_3 - A_2$$
$$y'' = 4 \cdot A_1 \cdot e^{2t} + \left(A_2 - 2 \cdot A_3 + A_3 \cdot t\right) \cdot e^{-t}$$
$$\text{and} \quad 14 = y''_0 = 4A_1 + A_2 - 2A_3$$

The system of equations.

$$\begin{pmatrix} 1 & 1 & 0 \\ 2 & -1 & 1 \\ 4 & 1 & -2 \end{pmatrix} \cdot \begin{pmatrix} A_1 \\ A_2 \\ A_3 \end{pmatrix} = \begin{pmatrix} 6 \\ -1 \\ 14 \end{pmatrix}$$

The solution: $A_1 = 2$, $A_2 = 4$, $A_3 = -1$ and $y = 2 \cdot e^{2t} + \left(4 - t\right) \cdot e^{-t}$. Once again, the determinant of the coefficient matrix of the IVP system is not zero (it's 9), so all possible IVP's can be solved with our general solution.

Here's an outrageous example meant to help us with all facets of the technique.

Problem. Describe the general solution to the DE.

$$(D - 2) \cdot D^4 \cdot (D^2 + 4D + 5)^2 \cdot (D^2 - 4)^2 \cdot (y'' - y) = 0$$

Solution. Writing $y'' - y = (D^2 - 1)y$, we get the operator polynomial; then we factor completely.

$$(D - 2) \cdot D^4 \cdot (D^2 + 4D + 5)^2 \cdot (D^2 - 4)^2 \cdot (D^2 - 1)$$
$$= (D - 2) \cdot D^4 \cdot (D^2 + 4D + 5)^2 \cdot (D - 2)^2 \cdot (D + 2)^2 \cdot (D - 1) \cdot (D + 1)$$
$$= (D - 2)^3 \cdot D^4 \cdot (D^2 + 4D + 5)^2 \cdot (D + 2)^2 \cdot (D - 1) \cdot (D + 1)$$

Note that all occurrences of a given root need to be put together. For instance, we need to collect the 3 occurrences of $D - 2$, since the exponent 3 tells us how many coefficients go with e^{2t} in the general solution.

We have left the factor $D^2 + 4D + 5$. The roots are available via the quadratic formula.

$$\frac{-4 \pm \sqrt{4^2 - 4 \cdot 5}}{2} = \frac{-4 \pm \sqrt{-4}}{2} = \frac{-4 \pm 2 \cdot i}{2} = -2 \pm i$$

Let's call these roots a, b for short. Now our factorization is complete.

$$(D - 2)^3 \cdot D^4 \cdot (D - a)^2 \cdot (D - b)^2 \cdot (D + 2)^2 \cdot (D - 1) \cdot (D + 1)$$

and our general solution is this.

$$\left(A_1 + A_2 t + A_3 t^2\right) \cdot \exp(2 \cdot t) + B_1 + B_2 t + B_3 t^2 + B_4 t^3$$
$$+ \left(C_1 + C_2 t\right) \cdot \exp(a \cdot t) + \left(E_1 + E_2 t\right) \cdot \exp(b \cdot t)$$
$$+ \left(F_1 + F_2 t\right) \cdot \exp(-2 \cdot t) + G_1 \cdot \exp(t) + H_1 \cdot \exp(-t)$$

Because a, b are complex, their exponentials involve cosine and sine. The arithmetic of an IVP here is easier if we just use a, b rather than the trig functions. It seems unlikely that we would do the arithmetic by hand, in any case. ■

Have you noticed that all our solutions are poly-exponential? This fact is the reason those functions were introduced in Chapter 1.

We want to move to application problems, and so we will skip over an important theoretical fact that will be proved later as Theorem 7.1 – that the operator polynomial technique gives the unique solution to each constant coefficient DE. This fact is crucial to applications, as we'll point out.

Let's begin with the homogeneous version of the equations that came up in the three examples at the beginning of the section. The equation (7.4) looks like this when its right side $g(t)$ is 0:

$$(7.9) \qquad\qquad y'' + p \cdot y' + q \cdot y = 0$$

where p, q are non-negative constants.

Problem. Let y be a solution to (7.9) when $p = 0$ and $q > 0$. Show that y oscillates. What is the frequency? (This case is called the *undamped* case.)

Solution. The equation is $(D^2 + q) \cdot y = 0$, and since $q > 0$, the roots are $\pm i\sqrt{q}$. Thus,[3]

$$y = A_1 \cdot \exp(i\sqrt{q} \cdot t) + A_2 \cdot \exp(-i\sqrt{q} \cdot t)$$

If you put y into cosine/sine form, you will see that the frequency is $\sqrt{q}/(2\pi)$.

Problem. Let y be a solution to (7.9) where p, q are positive. Show that $y \to 0$ as $t \to \infty$. (This is the *damped* case, and it contains three sub-cases, as we will see.)

---

[3]Here is where we use uniqueness. If the general solution were not unique, we would not know for certain what y looks like – the general solution would only give *one possible* form for y. Uniqueness guarantees that y *has* to have the form given here.

Solution. The operator polynomial is $D^2 + p \cdot D + q$. The quadratic formula gives the roots:

$$\frac{-p \pm \sqrt{p^2 - 4 \cdot q}}{2}$$

We break into cases depending on the sign of the *discriminant* $p^2 - 4q$.

Case 1. $p^2 - 4q > 0$. This case is called the *overdamped* case. Then the roots are distinct real numbers. Because the square root is positive and because $p > 0$, we see that

$$\frac{-p - \sqrt{p^2 - 4 \cdot q}}{2} < 0$$

For the other root, we need to be a little more careful. Since $q > 0$, we see that $p^2 - 4q < p^2$, and so $\sqrt{p^2 - 4q} < \sqrt{p^2} = p$. Thus,

$$-p + \sqrt{p^2 - 4q} < -p + p = 0 \quad \text{so that} \quad \frac{-p + \sqrt{p^2 - 4q}}{2} < 0$$

This proves that both roots are negative. Since this is all we need to know about the roots, we'll just call them α, β rather than use the more complicated formulas. Since α, β are distinct, we know that[4]

$$y = A_1 \cdot e^{\alpha \cdot t} + A_2 \cdot e^{\beta \cdot t}$$

for some constants A_1, A_2. Since α, β are negative, we see that $y \to 0$ as $t \to \infty$.

Case 2. $p^2 - 4q = 0$ This case is called the *critically damped* case. The operator polynomial has exactly one root: $-p/2$, and it is repeated. Then

$$y = \left(A_1 + A_2 \cdot t\right) \cdot \exp(-p \cdot t/2) = \frac{A_1 + A_2 \cdot t}{\exp(p \cdot t/2)}$$

for constants A_1, A_2. Again, $y \to 0$ as $t \to \infty$. (You might remember that this follows from L'Hôpital's Rule.)

Case 3. $p^2 - 4q < 0$. This case is called the *underdamped* case, and it is left to the problems at the end of the Chapter. ∎

---

[4]One last time: uniqueness allows us to assert what y *has* to look like.

There are many other applications in which constant coefficient problems occur. Our emphasis is on the operator method, and so we will continue to the case of non-homogeneous equations.

4. Non-homogeneous Problems

Now we are ready for the general DE of this chapter: $f(D) \cdot y = g(t)$ where $f(D)$ is an operator polynomial.

Let's start with an example: Solve the IVP

$$(7.10) \qquad (D^2 - 9) \cdot y = 10 \cdot e^{2t} \quad \text{and} \quad y_0 = -3, \; y_0' = 5$$

We know that $(D - 2) \cdot e^{2t} = 0$. Watch what happens when we apply $D - 2$ to both sides of the DE.

$$(D^2 - 9) \cdot y = 10 \cdot e^{2t}$$
$$(D - 2) \cdot (D^2 - 9) \cdot y = (D - 2) \cdot \left[10 \cdot e^{2t} \right]$$
$$(D - 2)(D - 3)(D + 3) \cdot y = 0$$

From our previous work, this tell us what y has to look like:

$$(7.11) \qquad y = A_1 \cdot e^{2t} + A_2 \cdot e^{3t} + A_3 \cdot e^{-3t}$$

An important point of logic: the equation (7.10) *led to* (7.11); it turns out that *not all solutions to* (7.11) are solutions to (7.10). We will discuss the reason for this in class. We call y in (7.11) the *over-general solution* to the non-homogeneous DE. For now, we take (7.11) and plug it back into (7.10) to get more information. In the following calculation, remember that $D^2 - 9$ wipes out e^{3t} and e^{-3t}. It will be easier to deal just with the left side of the

DE and bring the right side in later.

$$(D^2 - 9) \cdot \left[A_1 \cdot e^{2t} + A_2 \cdot e^{3t} + A_3 \cdot e^{-3t} \right]$$
$$= (D^2 - 9) \cdot \left[A_1 \cdot e^{2t} \right] + (D^2 - 9) \cdot \left[A_2 \cdot e^{3t} \right] + (D^2 - 9) \cdot \left[A_3 \cdot e^{-3t} \right]$$
$$= D^2 \cdot \left[A_1 \cdot e^{2t} \right] - 9 \cdot \left[A_1 \cdot e^{2t} \right] + 0 + 0$$
$$= 4 \cdot A_1 \cdot e^{2t} - 9 \cdot A_1 \cdot e^{2t}$$
$$= -5 \cdot A_1 \cdot e^{2t}$$

The right side of the DE was $10 \cdot e^{2t}$. To get $(D^2 - 9)y = 10 \cdot e^{2t}$, we need $A_1 = -2$. Notice that the constants A_2, A_3 drop out. That says that they can be anything! So here is the general solution to (7.10).

$$(7.12) \qquad\qquad y = -2 \cdot e^{2t} + A_2 \cdot e^{3t} + A_3 \cdot e^{-3t}$$

Since every pair of values A_2, A_3 in (7.12) gives us a solution to the DE, we see that $y = -2 \cdot e^{2t}$ is a solution (with $A_2 = A_3 = 0$). The function $-2 \cdot e^{2t}$ is called a *particular solution*. The general solution y consists of the particular solution along with $A_2 \cdot e^{3t} + A_3 \cdot e^{-3t}$, which comes from the homogeneous equation $(D^2 - 9) \cdot y = 0$. This homogeneous equation is called the *related homogeneous equation*.

Finally we consider the initial conditions to (7.10): $y_0 = -3$, $y_0' = 5$. Taking the solution (7.12) we compute

$$y = -2 \cdot e^{2t} + A_2 \cdot e^{3t} + A_3 \cdot e^{-3t} \quad \text{and} \quad -3 = y_0 = -2 + A_2 + A_3$$
$$y' = -4 \cdot e^{2t} + 3 \cdot A_2 \cdot e^{3t} - 3 \cdot A_3 \cdot e^{-3t} \quad \text{and} \quad 5 = y_0' = -4 + 3 \cdot A_2 - 3 \cdot A_3$$

The system of equations:

$$\begin{pmatrix} 1 & 1 \\ 3 & -3 \end{pmatrix} \cdot \begin{pmatrix} A_2 \\ A_3 \end{pmatrix} = \begin{pmatrix} -1 \\ 9 \end{pmatrix}$$

This has solution $A_2 = 1$, $A_3 = -2$, so that $y = -2 \cdot e^{2t} + e^{3t} - 2 \cdot e^{-3t}$.

As in the case of homogeneous equations, notice that the coefficient matrix for the undetermined constants has non-zero determinant: we would be able to solve all possible IVP's.

Problem. Solve

$$y'' + 16 \cdot y = 25 \cdot e^{3t} \quad \text{and} \quad y_0 = 5, \ y_0' = 4$$

Solution.   The equation is

(7.13) $(D^2 + 16) \cdot y = 25 \cdot e^{3t}$

The related homogeneous equation is $(D^2 + 16) \cdot y = 0$, and $D^2 + 16 = (D - 4i)(D + 4i)$. We can use $D - 3$ to make the right side go away.

$$(D^2 + 16) \cdot y = 25 \cdot e^{3t}$$

$$(D - 3) \cdot (D^2 + 16) \cdot y = (D - 3) \cdot \left[25 \cdot e^{3t} \right]$$

$$(D - 3) \cdot (D - 4i) \cdot (D + 4i) \cdot y = 0$$

This shows that y looks like this:

$$y = A_1 \cdot e^{3t} + A_2 \cdot e^{4it} + A_3 \cdot e^{-4it}$$

We need to plug this back into the original equation (7.13). As above, we neglect the right side for now.

$$(D^2 + 16) \cdot \left[A_1 \cdot e^{3t} + A_2 \cdot e^{4it} + A_3 \cdot e^{-4it} \right]$$

$$= (D^2 + 16) \cdot \left[A_1 \cdot e^{3t} \right] + (D^2 + 16) \cdot \left[A_2 \cdot e^{4it} + A_3 \cdot e^{-4it} \right]$$

$$= (D^2 + 16) \cdot \left[A_1 \cdot e^{3t} \right] + 0$$

$$= D^2 \cdot \left[A_1 \cdot e^{3t} \right] + 16 \cdot \left[A_1 \cdot e^{3y} \right]$$

$$= 9 \cdot A_1 \cdot e^{3t} + 16 \cdot \left[A_1 \cdot e^{3y} \right]$$

$$= 25 \cdot A_1 \cdot e^{3t}$$

Bringing in the right side we have

$$25 \cdot A_1 \cdot e^{3t} = 25 \cdot e^{3t}$$

We see that $A_1 = 1$ and A_2, A_3 are arbitrary, and so our solution is

$$y = e^{3t} + A_2 \cdot e^{4it} + A_3 \cdot e^{-4it}$$

The initial conditions:

$$y = e^{3t} + A_2 \cdot e^{4it} + A_3 \cdot e^{-4it} \quad \text{and} \quad 5 = y_0 = 1 + A_2 + A_3$$
$$y' = 3 \cdot e^{3t} + 4i \cdot A_2 \cdot e^{4it} - 4i \cdot A_3 \cdot e^{-4it} \quad \text{and} \quad 4 = y_0' = 3 + 4i \cdot A_2 - 4i \cdot A_3$$

The system of equations.

$$\begin{pmatrix} 1 & 1 \\ 4i & -4i \end{pmatrix} \cdot \begin{pmatrix} A_2 \\ A_3 \end{pmatrix} = \begin{pmatrix} 4 \\ 1 \end{pmatrix}$$

Solution: $A_2 = 2 - (i/8)$, $A_3 = 2 + (i/8)$, so that

$$y = e^{3t} + \left(2 - \frac{i}{8} \right) \cdot e^{4it} + \left(2 + \frac{i}{8} \right) \cdot e^{-4it}$$

The next example is uglier. When there is more than one function on the right side, we need more than one factor to get rid of it. Also, a repeated root may arise.

Problem.   Solve

$$y' + 5 \cdot y = t \cdot e^{3t} + e^{-5t} \quad \text{and} \quad y_0 = 0$$

Solution.   (By the way: this is first order linear, so we could use the technique of Chapter 2 to check our work here.) The left side is $(D + 5) \cdot y$. We need $(D - 3)^2$ to get rid of $t \cdot e^{3t}$ and $D + 5$ to get rid of e^{-5t}. So, we apply $(D - 3)^2 \cdot (D + 5)$ to both sides of the DE.

$$(D - 3)^2 \cdot (D + 5) \cdot (D + 5) \cdot y = (D - 3)^2 \cdot (D + 5) \cdot \left[t \cdot e^{3t} + e^{-5t} \right]$$
$$(D - 3)^2 \cdot (D + 5)^2 \cdot y = 0$$

Then

$$y = \left(A_1 + A_2 \cdot t\right) \cdot e^{3t} + \left(A_3 + A_4 \cdot t\right) \cdot e^{-5t}$$

The homogeneous part of the solution is somewhat hidden. Recall that the left side of the original DE is $D + 5$; this clears $A_3 \cdot e^{-5t}$. We put y into the original left side; we'll bring the right side in momentarily. The derivatives take a while to resolve – you might want to check our work.

$$(D + 5) \cdot \left[\left(A_1 + A_2 \cdot t\right) \cdot e^{3t} + \left(A_3 + A_4 \cdot t\right) \cdot e^{-5t}\right]$$
$$= (D + 5) \cdot \left[\left(A_1 + A_2 \cdot t\right) \cdot e^{3t}\right] + (D + 5) \cdot \left[\left(A_3 + A_4 \cdot t\right) \cdot e^{-5t}\right]$$
$$= D \cdot \left[\left(A_1 + A_2 \cdot t\right) \cdot e^{3t}\right] + 5 \cdot \left(A_1 + A_2 \cdot t\right) \cdot e^{3t}$$
$$\quad + D \cdot \left[\left(A_3 + A_4 \cdot t\right) \cdot e^{-5t}\right] + 5 \cdot \left(A_3 + A_4 \cdot t\right) \cdot e^{-5t}$$
$$= \left(8 \cdot A_1 + A_2 + 8 \cdot A_2 \cdot t\right) \cdot e^{3t} + A_4 \cdot e^{-5t}$$

Bringing in the right side of the DE:

$$\left(8 \cdot A_1 + A_2 + 8 \cdot A_2 \cdot t\right) \cdot e^{3t} + A_4 \cdot e^{-5t} = t \cdot e^{3t} + e^{-5t}$$

Matching up like terms, we see that

$$8A_1 + A_2 = 0, \quad 8A_2 = 1, \quad A_4 = 1$$

Then $A_2 = 1/8$ and $A_1 = -1/64$. The constant A_3 is arbitrary, and so our general solution is

$$y = \left(-\frac{1}{64} + \frac{t}{8}\right) \cdot e^{3t} + \left(A_3 + t\right) \cdot e^{-5t}$$

Finally, the initial condition $y_0 = 0$ shows that $A_3 = 1/64$.

What about a trigonometric function on the right side? The next problem demonstrates that the calculations necessary can be quite complicated; at least those calculations are straightforward!

Problem. Solve the IVP

$$y' - 3 \cdot y = -169 \cdot t \cdot \sin(2t) \quad \text{and} \quad y_0 = 2$$

Solution. As in our work on homogeneous equations, we know that $\sin(2t)$ can be written in terms of e^{2it} and e^{-2it}. Thus, to get rid of it, we need $(D - 2i)(D + 2i) = D^2 + 4$. The presence of the factor t in $t \cdot \sin(2t)$ shows that there is repetition of these roots, and we need $(D^2 + 4)^2$ to get rid of $t \cdot \sin(2t)$. We have

$$(D^2 + 4)^2 \cdot (D - 3) \cdot y = 0$$

and so

$$y = (A_1 + A_2 \cdot t) \cdot \cos(2t) + (A_3 + A_4 \cdot t) \cdot \sin(2t) + A_5 \cdot e^{3t}$$

You might wonder whether we really need the cosine terms; in fact, you might wonder whether we need the A_3 term since the right side of the original DE involves only $t \cdot \sin(2t)$. As we will see, *all* the terms are needed. Next step: plug this form into the original DE. Since e^{3t} is canceled by the left side operator $D - 3$, we can ignore that term. It takes a while to apply $D - 3$ to the other terms. Here is the result, equated to the right side of the original DE.

$$\left[-3A_1 + A_2 + 2A_3 + t \cdot \left(-3A_2 + 2A_4 \right) \right] \cdot \cos(2 \cdot t)$$
$$+ \left[-2A_1 - 3A_3 + A_4 + t \cdot \left(-2A_2 - 3A_4 \right) \right] \cdot \sin(2 \cdot t)$$
$$= \left[0 + 0 \cdot t \right] \cdot \cos(2t) + \left[0 - 169 \cdot t \right] \cdot \sin(2t)$$

Equating like terms, we obtain the following system of equations.

$$\begin{pmatrix} -3 & 1 & 2 & 0 \\ 0 & -3 & 0 & 2 \\ -2 & 0 & -3 & 1 \\ 0 & -2 & 0 & -3 \end{pmatrix} = \cdot \begin{pmatrix} A_1 \\ A_2 \\ A_3 \\ A_4 \end{pmatrix} = \begin{pmatrix} 0 \\ 0 \\ 0 \\ -169 \end{pmatrix}$$

This has solution $A_1 = 12$, $A_2 = 26$, $A_3 = 5$, $A_4 = 39$. Thus, the general solution to the DE looks like this.

$$(12 + 26 \cdot t) \cdot \cos(2t) + (5 + 39 \cdot t) \cdot \sin(2t) + A_5 \cdot e^{3t}$$

Finally, the initial condition $y_0 = 2$ gives $A_5 = -10$, and we are done!

We will do additional examples in class to make sure you understand the pattern of the solution.

5. Examples and Applications

In the RLC circuit considered at the beginning of this chapter, the right side of the DE (7.1) is the derivative of the potential V'. When the current is alternating, V' is typically expressed as a sum of signals, each with its own amplitude and frequency. It is common for the frequencies to be multiples of some base frequency, and so here is what V' looks like.

$$(7.14) \qquad V' = \sum_{k=0}^{n} \Big[a_k \cdot \cos(k \cdot w \cdot t) + b_k \cdot \sin(k \cdot w \cdot t) \Big]$$

where all the a_k and b_k are constants.[5] You might like to observe that the base frequency is $w/(2\pi)$.

We also considered an example of a mechanical system – a shock absorber. The right side of the DE (7.3) has the form $-g + k \cdot (f(t) + L)$ for some function f. It is interesting to think about the response to an oscillating terrain. In this case, $f(t)$ would be written as in (7.14), and so the DE (7.3) can be very similar to the DE (7.1) for the circuit. Here is an analysis that pertains to both.

---

[5]We have included $k = 0$ and $\sin(0wt) = 0$ wipes out b_0. It would be typical for V' that $a_0 = 0$ as well, but we will keep the general form of the equation here, since it is useful in other problems.

Response to a signal. We have seen that DE's of the following form arise in several contexts.

$$(7.15) \qquad y'' + p \cdot y' + q \cdot y = f(t)$$

where p, q are positive constants.[6] We want to think of $f(t)$ as an external *signal* to which the variable y *responds*. One of the most common ways to write a signal is as a sum of signals of given frequencies. Equation (7.14) does this. As we pointed out, the number w corresponds to a *fundamental frequency* $w/(2\pi)$. The other frequencies present are multiples of this one.

We want to see that the response y is the sum of responses to the individual frequencies: let y_k be a solution to this DE:

$$y_k'' + p \cdot y_k' + q \cdot y_k = a_k \cdot \cos(k \cdot w \cdot t) + b_k \cdot \sin(k \cdot w \cdot t)$$

Then the sum y of the y_k is a solution to the entire DE (7.15). Watch.

$$y'' + p \cdot y' + q \cdot y = \left(\sum_{k=0}^{n} y_k \right)'' + p \cdot \left(\sum_{k=0}^{n} y_k \right)' + q \cdot \left(\sum_{k=0}^{n} y_k \right)$$

$$= \sum_{k=0}^{n} y_k'' + p \cdot \sum_{k=0}^{n} y_k' + q \cdot \sum_{k=0}^{n} y_k$$

$$= \sum_{k=0}^{n} \left(y_k'' + p \cdot y_k' + q \cdot y_k \right)$$

$$= \sum_{k=0}^{n} \left[a_k \cdot \cos(k \cdot w \cdot t) + b_k \cdot \sin(k \cdot w \cdot t) \right]$$

$$= f(t)$$

Do you see that this is a kind of superposition, except that each y_k brings its own right side along with it.

We can use the sum of the y_k's to get a particular solution to (7.15), and then the general solution is the sum of the particular solution and a solution

---

[6]This is the *damped* case with its three sub-cases.

to the related homogeneous equation

$$y'' + p \cdot y' + q \cdot y = 0$$

When we analyzed the damped case, we saw that solutions to the homogeneous equation go to 0 as $t \to \infty$. These solutions are called *transient solutions*, because they disappear after a while. Thus, over time, y responds to the input signal as the sum of the y_k.

To solve for a single y_k, we use the operator polynomial:

$$(7.16) \qquad (D^2 + p \cdot D + q) \cdot y_k = a_k \cdot \cos(k \cdot w \cdot t) + b_k \cdot \sin(k \cdot w \cdot t)$$

Recall from the complex exponential function, that $\cos(kwt)$ and $\sin(kwt)$ involve exponentials in $\pm ikwt$. Thus, the operator that gets rid of both cosine and sine is

$$(D - ikw) \cdot (D + ikw) = D^2 + k^2 w^2$$

Going back to (7.16), we have

$$(D^2 + k^2 w^2) \cdot (D^2 + p \cdot D + q) \cdot y_k = 0$$

It is important to note that these two factors have no roots in common, so that there are no repeated roots here. This is because the roots of $D^2 + k^2 w^2$ are $\pm ikw$ and the roots of the other factor have real part $-p/2 \neq 0$. Since there are no repeated roots, we can find a particular solution without a homogeneous part; we write this solution in cosine/sine form.

$$y_k = A_k \cdot \cos(kwt) + B_k \cdot \sin(kwt)$$

Plugging this back into (7.16) and working out the details, we obtain

$$\left(-k^2 w^2 A_k + pkw B_k + qA_k \right) \cdot \cos(kwt)$$
$$+ \left(-k^2 w^2 B_k - pkw A_k + qB_k \right) \cdot \sin(kwt)$$
$$= a_k \cdot \cos(kwt) + b_k \cdot \sin(kwt)$$

When we write out the equations that A_k, B_k have to satisfy, we get the following system of linear equations.

(7.17)
$$\begin{pmatrix} q - k^2w^2 & pkw \\ -pkw & q - k^2w^2 \end{pmatrix} \cdot \begin{pmatrix} A_k \\ B_k \end{pmatrix} = \begin{pmatrix} a_k \\ b_k \end{pmatrix}$$

The solution is messy!

$$A_k = \frac{(q - k^2w^2) \cdot a_k - pkw \cdot b_k}{(q - k^2w^2)^2 + p^2k^2w^2} \quad \text{and} \quad B_k = \frac{pkw \cdot a_k + (q - k^2w^2) \cdot b_k}{(q - k^2w^2)^2 + p^2k^2w^2}$$

One piece of very significant information is the ratio of amplitudes of the responding signal with the input signal. This ratio can be computed from the solution, and it simplifies nicely.

(7.18)
$$\frac{\sqrt{A_k^2 + B_k^2}}{\sqrt{a_k^2 + b_k^2}} = \frac{1}{\sqrt{(q - k^2w^2)^2 + p^2k^2w^2}}$$

This ratio of amplitudes is called the *gain* of the response.

In both mechanical systems and in electrical circuits, the coefficient c can be adjustable: a spring mechanism in a mechanical system can have an adjustable spring constant, and the capacitance in an electrical circuit can be adjustable. The adjustment allows us to *tune* the response toward a given frequency. Rather than go through an algebraic analysis, we give a representative example. Here is the graph of the gain (7.18) when $c = 9$ and $b = 1/10$ and $w = 1$. The horizontal axis shows k, which indexes the various input signals; the vertical axis shows the gain.

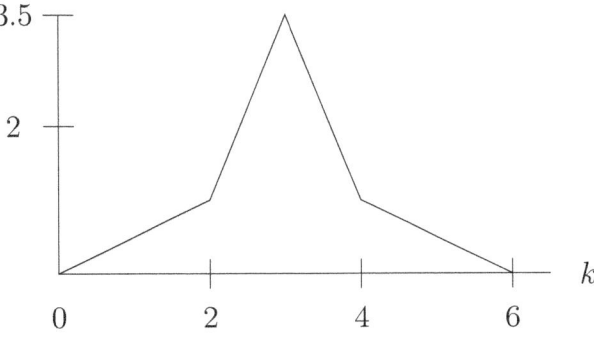

We observe a gain spike when $k = 3$. Thus, y_3 has a relatively larger amplitude than the other y_k. The mechanical system or circuit is tuned to that frequency; we might say that the system *detects* that frequency. In RLC circuits, this is the principle of tuning to radio signal frequency; a radio tuner is an adjustable capacitor. In mechanical systems, this is the principle of *resonance*. Resonance can be good: our mechanical system might be intended to detect vibrations at a given frequency and to respond to those vibrations. Resonance can be bad: large amplitude response can cause the mechanical system to break. There are several disasters involving bridges and other structures that have been attributed to resonance. See [**2**, p.184]

The sliding block.   We will discuss this problem in class; you might want to look it over in advance, since there are several DE's of the kind already discussed that you should be able to solve. The twist in this problem – a discontinuous right side, makes it interesting. We have a block of mass m on a spring along the x-axis. The spring constant is k. There is a constant force of friction, constant in magnitude and *against* the direction of motion. Thus, our system is

$$m \cdot x'' + k \cdot x = -F \qquad\qquad \text{when } x' > 0$$
$$m \cdot x'' + k \cdot x = F \qquad\qquad \text{when } x' < 0$$

There is a *jump* in the driving force when $x' = 0$. For convenience, assume that $m = 1$; write $w = \sqrt{k}$.

Case 1.   Assume we are given x_0 and $x_0' = 0$ and $x_0'' > 0$ so that x' should increase and become positive. You should be able to solve $x'' + kx = -F$ to get

$$x = \left(x_0 + \frac{F}{k} \right) \cdot \cos(w \cdot t) - \frac{F}{k}$$

We have

$$x' = -w \cdot \left(x_0 + \frac{F}{k}\right) \cdot \sin(w \cdot t)$$

and so in order to have $x' > 0$ as $t > 0$, we need

$$x_0 < -\frac{F}{k}$$

(The block must be far enough to the left that the spring pulls harder than friction. Note that $x_0'' = -w^2(x_0 + F/k) > 0$.) This system breaks down when $x' = 0$ again, and this happens at $t = \pi/w$. Notice that

$$x(\pi/w) = -x_0 - 2 \cdot \frac{F}{k}$$

Case 2. We are given x_0 and $x_0' = 0$ and $x'' < 0$ so that x' should decrease and become negative. Then you should be able to solve $x'' + kx = F$

$$x = \left(x_0 - \frac{F}{k}\right) \cdot \cos(w \cdot t) + \frac{F}{k}$$

with

$$x' = -w \cdot \left(x_0 - \frac{F}{k}\right) \cdot \sin(w \cdot t)$$

We need $x' < 0$ for $t > 0$ and so

$$x_0 > \frac{F}{k}$$

and again the interpretation is obvious: spring pulls harder than friction. (Note that $x_0'' = -w^2(x_0 - F/k) < 0$.) We have $x' = 0$ again at $t = \pi/w$ and

$$x(\pi/w) = -x_0 + 2 \cdot \frac{F}{k}$$

Here is a problem we will discuss in class – you might give it a try, taking care to notice which case (of the two discussed above) you are in at a given time.

Problem.   Suppose we start with $x_0' = 0$ and

$$\frac{7F}{k} < x_0 \leq \frac{9F}{k}$$

What happens to the block? ∎

6. Existence and Uniqueness of Solutions

Uniqueness of solution is crucial to applications, for it tells us what the solution to an applied problem *has* to look like – it gives us information with logical force behind it. In the following theorem, we prove uniqueness for constant coefficient IVP's when the right side is poly-exponential. We also show that the operator technique finds a solution to every possible such IVP, so that the technique is universal. The proof is complicated, and you will probably find it hard to follow. It is given here for the sake of completeness, and especially for those interested in mathematical detail. Unfortunately, we will not have much time to discuss it in class.

THEOREM 7.1. *Let $f(D)$ be a monic[7] operator polynomial of degree $n \geq 1$, and let $g(t)$ be a poly-exponential. Consider the IVP consisting of the DE: $f(D) \cdot y = g(t)$ and given values $y_0^{(k)}$ for $k = 0, 1, \ldots, n - 1$. Then this IVP has a unique solution, and the unique solution is a poly-exponential function.*

PROOF. It will be necessary to separate the initial values from the notation for the solution y. Write v_k for the desired initial value of the k-th derivative of the solution. Then the IVP we are trying to solve asks to have $f(D) \cdot y = g(t)$ and $y_0^{(k)} = v_k$ for $0 \leq k \leq n - 1$.

We use induction on the degree n. If $n = 1$, then $f(D) = D - c$ for some number c, and the equation $(D - c)y = g(t)$ is first order linear. Theorem 2.1 in Chapter 2 tells us that the IVP with $y_0 = v_0$ has a unique solution and that this solution is poly-exponential, and so we are done in this case.

---

[7]The word *monic* means that the leading coefficient of f is 1.

Assume that the present Theorem is true when the degree of the operator polynomial is $n-1$, and assume that the polynomial $f(D)$ has degree n. That polynomial at least one complex number root.[8] Say $f(\beta) = 0$, so that $f(D) = (D - \beta) \cdot h(D)$ where $h(D)$ is a monic operator polynomial of degree $n-1$.

We consider the following IVP:

$$(7.19) \qquad h(D) \cdot u = g(t) \quad \text{and} \quad u_0^{(k)} = v_{k+1} - \beta \cdot v_k \quad \text{for} \quad 0 \le k \le n-2$$

Since k runs from 0 to $n-2$, the number $k+1$ goes from 1 to $n-1$, and so the values v_{k+1} are defined. The initial conditions have been cleverly chosen; you'll see.

The polynomial $h(D)$ has degree $n-1$, and so induction tells us that (7.19), has a unique solution u, and this unique solution is poly-exponential. Now we can define y. We use Theorem 2.1 in Chapter 2 again: the IVP $(D - \beta) \cdot y = u$, and $y_0 = v_0$ has a unique solution, and that solution is poly-exponential.

We claim that y is a solution to the original IVP, and that it is the unique solution. First: it's a solution! Compute

$$f(D) \cdot y = h(D) \cdot (D - \beta) \cdot y = h(D) \cdot u = g(t)$$

and we see that y satisfies the DE. Initial values? We have already specified that $y_0 = v_0$. We have $y' - \beta \cdot y = u$. Putting $t = 0$, we have

$$y_0' - \beta \cdot y_0 = u_0 = v_1 - \beta \cdot v_0$$

and since $y_0 = v_0$, we see that $y_0' = v_1$. Taking the derivative of the equation $y' - \beta \cdot y = u$, and taking $t = 0$, we get

$$y_0'' - \beta \cdot y_0' = u_0' = v_2 - \beta \cdot v_1$$

---

[8]The *Fundamental Theorem of Algebra* is the very important fact that every polynomial of degree at least 1 with complex number coefficients has at least one complex number root. Even though there are calculus-level proofs of this theorem, it is rare for a proof to be presented below the advanced level, and we will not present a proof here.

Since $y_0' = v_1$, we see that $y_0'' = v_2$. We can keep going, and we will see that $y_0^{(k)} = v_k$ for each $k = 0, 1, \ldots, n-1$. Thus y satisfies the IVP.

It remains to show that y is unique. If y is a solution to the IVP, then we can define $u = (D - \beta) \cdot y$. Exactly as above, this leads to $u_0^{(k)} = v_{k+1} - \beta \cdot v_k$ for $0 \le k \le n-2$ (except that now we are reasoning from the initial values of y to those of u rather than from u to y as before). Then

$$h(D) \cdot u = h(D) \cdot (D - \beta) \cdot y = f(D) \cdot y = g(t)$$

so that u is a solution to $h(D) \cdot u = g(t)$. The initial values of u have already been determined to be as in (7.19), and we see that u is the function called u we defined before. Since y was unique, given u, this proves that y is unique. $\qquad\square$

7. Problems

7.1. Solve the following IVP's.

    a) $y'' + 2 \cdot y' - 15 \cdot y = 0, \quad y_0 = 5, \ y_0' = 31$

    b) $y'' + 9 \cdot y = 0, \quad y_0 = 3, \ y_0' = 6$

    c) $y^{(3)} - y'' - 2 \cdot y' = 0, \quad y_0 = 8, \ y_0' = 2, \ y_0'' = 4$

7.2. Find the general solution to the following DE's. If there are non-real roots involved, write the answer both in complex exponential form and in sine/cosine form.

    a) $y^{(3)} + y'' + 16 \cdot y' + 16 \cdot y = 0$

    b) $(D - 2)^2 \cdot (D + 5)^3 \cdot \left(y^{(3)} + 4 \cdot y'' - 5 \cdot y'\right) = 0$

7.3. Solve the following IVP's.

      a) $y' - 7 \cdot y = 10 \cdot e^{2t} + 8 \cdot e^{-t}, \quad y_0 = 4$

      b) $y'' - 2 \cdot y' - 8 \cdot y = 6 \cdot e^{4t}, \quad y_0 = -2, \ y_0' = -1$

      c) $y'' - 5 \cdot y' + 4 \cdot y = 3 \cdot \sin(t) - 5 \cdot \cos(t), \quad y_0 = 0, \ y_0' = 4$

7.4. Find the form of the solution to the following DE, and identify the homogeneous part of the solution: $y^{(5)} + 2 \cdot y^{(3)} = \sin(\sqrt{2} \cdot t) + 4t - e^t$.

7.5. In the underdamped case of the mechanical system, show that $y \to 0$ as $t \to \infty$. Find the frequency with which y oscillates as it goes to 0.

7.6. The *most famous DE* is $y'' + y = 0$. Solve this equation using the technique of this section, and put the general solution into cosine/sine form.

7.7. The pendulum clock model is Example 3 at the beginning of this chapter. Find the length L in feet given that $g \approx 32$ ft/sec^2, $k \approx 5$, and that the DE is underdamped with period 2 seconds. (This corresponds to a design of Huygens around 1656.[**10**, p.330ff])

7.8. A floating buoy experiences an upward acceleration equal to 11 times the length of its submerged part[9] and a downward constant acceleration of magnitude 9.8 due to gravity. (Acceleration uses length in meters and time in seconds.) Write down the DE governing the submerged length L and find its general solution. (Hint: use the water surface as origin and measure L downward. Up is negative!)

---

[9]Technical simplification: we're assuming that the mass density of the buoy is 1/11 that of water.

7.9. (This problem will be used in Chapter 14.) Suppose that $x(t)$ is defined for $0 \le t \le 1$ and $x'' + k \cdot x = 0$ there. Assume also that $x(0) = 0$ and $x(1) = 0$ but that $x(t) \ne 0$ for some t between 0 and 1. Show that $k = \pi^2 \cdot n^2$ for some integer n. (Hint: if $k < 0$, show that the solution cannot satisfy the conditions.)

7.10. (*Reduction of Order*[10]) We consider the equation

(7.20) $$y'' + p(t) \cdot y' + q(t) \cdot y = g(t)$$

(Since p, q are not constants, the D operator won't help.) Suppose we can find a particular solution $u(t)$ to the related homogeneous equation

$$u'' + p(t) \cdot u' + q(t) \cdot u = 0$$

Set $y = u \cdot v$ where v is an unknown function of t; substitute into (7.20) and show that you obtain a first order linear equation in v'.

7.11. Use the technique of the previous problem to solve the DE

$$y'' - \frac{2}{t^2} \cdot y = t^{5/3}$$

given that $u = 1/t$ is a solution to $u'' - (2/t^2) \cdot u = 0$. (Note: don't worry about initial conditions, and feel free to choose particular constants of integration when you need them.)

7.12. Consider the DE: $t^2 \cdot y'' + 2 \cdot t \cdot y' - 12 \cdot y = 0$. Show that there is a solution of the form $y = t^n$. (Hint: Note that this is **not** a constant coefficient problem, so the operator D is not relevant. Plug in t^n as solution and solve for n.) This equation is called an *Euler-Cauchy DE*[11]

---

[10]This technique occurs in a paper Euler wrote when he was 21 years old. An excerpt of an English translation of that paper occurs in [**6**, p.638].

[11]I have not been able to find the origin of this equation. Euler and Cauchy lived in different centuries, so it must be the case that each of them considered such an equation independently.

7.13. Many classes of polynomials that occur in applied problems are defined by DE's. Here are some famous examples; in each case, the subscript n labels a polynomial of degree n. The initial values are different than what we have seen, but they define unique polynomials in each case.

(a) *Chebyshev polynomials*[12] $(1 - x^2) \cdot T_n''(x) - x \cdot T_n'(x) + n^2 \cdot T_n(x) = 0$ and $T_n(1) = 1$.

(b) *Hermite polynomials* $H_n''(x) - 2 \cdot x \cdot H_n'(x) + 2 \cdot n \cdot H_n(x) = 0$ and the leading coefficient of $H_n(x)$ is 2^n.

(c) *Legendre polynomials* $(1 - x^2) \cdot L_n''(x) - 2 \cdot x \cdot L_n'(x) + n \cdot (n+1) \cdot P_n(x) = 0$ and $P_n(1) = 1$.

Find $T_3(x)$ and $H_3(x)$ and $L_3(x)$. (Hint: the D operator is not helpful; in each case you are solving linear equations in the coefficients of the polynomials.)

---

[12]You might find it interesting that $T_n(\cos(\theta)) = \cos(n\theta)$ for each n.

CHAPTER 8

Vector Spaces

1. Fundamental Vector Spaces

Vectors in the plane, vectors in space, polynomials, functions, and matrices: all of these can be added, each to each. These objects can also be multiplied by scalars. Furthermore, the properties of these operations are similar in each case – that's why the word *addition* is used in each case. These sets are examples of what are called *vector spaces* – the name for sets that have an addition and scalar multiplication *similar* to the addition and scalar multiplication of space vectors and matrices.

We want to answer these questions: 1) What does the word *similar* mean, as used just now? 2) Why would we want to study a large class of *similar* sets? (Won't a level of abstraction obscure the important features of individual cases?)

The second question first: Vector spaces are ubiquitous in applications, and although they vary from example to example there is a small, but very useful, list of facts that are true about all vector spaces. If we once establish those facts generally (abstractly), then in each individual instance we will automatically have those facts at our disposal with no special argument to give in the various special cases. This actually *frees* us to study the peculiar properties of an individual example without digressing to "re-invent the vector-space wheel."

As to the first question, we list the common properties of addition and scalar multiplication as they occur in the following examples:

(1) $\mathbb{R}^n$ for each positive integer n;

(2) the $m \times n$ matrices for each pair m, n of positive integers;

(3) the set of polynomials;

(4) the set of differentiable functions defined on the real numbers.

There are many other examples that can be given; these seem basic.

Here are the properties, labeled V1, V2, and so on. We have a set V; for each u, v in V, their sum $u + v$ is defined as an element of V; for each u in V and each real number α, the product $\alpha \cdot u$ is defined as an element of V. Furthermore, we have the following.

V1: $u + v = v + u$ for all u, v in V;

V2: $(u + v) + w = u + (v + w)$ for all u, v, w in V;

V3: there is a zero element $\mathbb{O}$ in V such that $u + \mathbb{O} = u$ for all u in V;

V4: $(\alpha + \beta) \cdot u = (\alpha \cdot u) + (\beta \cdot u)$ for all numbers α, β and all u in V;

V5: $\alpha \cdot (u + v) = (\alpha \cdot u) + (\alpha \cdot v)$ for all u, v in V and all numbers α;

V6: $\alpha \cdot (\beta \cdot u) = (\alpha \cdot \beta) \cdot u$ for all numbers α, β and u in V;

V7: $0 \cdot u = \mathbb{O}$ and $1 \cdot u = u$, for all u in V.

It is easy to see that these properties hold in the examples we gave. For instance, when the set V is $\mathbb{R}^n$, the zero element $\mathbb{O}$ is the zero vector: $\mathbb{O}_{n \times 1}$. When V is the set of polynomials, $\mathbb{O}$ is the zero polynomial 0. The particular $\mathbb{O}$ will vary from example to example; property (V3) shows how this zero element must behave: it does nothing when added to other elements.

The properties listed hold in many, many other settings. The exercises give a couple more examples. To play with these properties a little, notice that each element u of the set V has a *negative* $(-1) \cdot u$ (denoted $-u$, as usual). The negative of a vector is its additive inverse, as is demonstrated by the following calculation, using (V7), (V4), and (V7) again.

$$u + (-1) \cdot u = 1 \cdot u + (-1) \cdot u = (1 - 1) \cdot u = 0 \cdot u = \mathbb{O}$$

When the set V has these properties, it is called a *vector space*. The elements of a vector space are often called *vectors*. You may have thought of vectors as directed line segments (drawn, perhaps, with an arrow on one end to indicate the direction). We are proposing to use the word *vector* much more generally. Thus, a *vector* could be a function or a matrix, for instance. Also, the meaning of *zero* depends on the vector space: when the "vectors" are functions, the "zero vector" is the zero function; when the "vectors" are 2×3 matrices, the "zero vector" is $\mathbb{O}_{2\times3}$. We remind you that the use of the word vector in so many different contexts is meant to create a maximum level of efficiency in applications.[1]

2. Finite-Dimensional Subspaces

How do vector spaces arise in practice? Equivalently: how do we recognize that we have a vector space? We will give the answer and then explain the terms involved: a vector space occurs as the *span* of a set of vectors in some previously given vector space.

Let's start with an example. Let A be an $m \times n$ matrix. The set of solutions to the system of equations $AX = \mathbb{O}_{m\times1}$ is called the *null space* of A. (The term null space is related to the nullity in a way that we will soon see.) We want to look at the solutions to such an equation in a different way than we have in the past. To give a specific example, suppose we have a 3×5 matrix

---

[1]When we defined the matrix operations, we called attention to the different uses of the plus and times signs. For a given vector space, those signs have to be defined carefully in the particular context.

A, and Elimination puts it into the following row-echelon form.

$$\begin{pmatrix} 1 & 0 & -2 & 0 & 4 \\ 0 & 1 & -1 & 3 & 6 \\ 0 & 0 & 0 & 0 & 0 \end{pmatrix}$$

Then the null space of A is the set of vectors defined by the three free variables x_3, x_4, x_5. We write the solution in a way that isolates the free variables.

$$(8.1) \qquad \begin{pmatrix} 2x_3 - 4x_5 \\ x_3 - 3x_4 - 6x_5 \\ x_3 \\ x_4 \\ x_5 \end{pmatrix} = x_3 \cdot \begin{pmatrix} 2 \\ 1 \\ 1 \\ 0 \\ 0 \end{pmatrix} + x_4 \cdot \begin{pmatrix} 0 \\ -3 \\ 0 \\ 1 \\ 0 \end{pmatrix} + x_5 \cdot \begin{pmatrix} -4 \\ -6 \\ 0 \\ 0 \\ 1 \end{pmatrix}$$

The form on the right is said to be a *linear combination* of the matrices

$$(8.2) \qquad \begin{pmatrix} 2 \\ 1 \\ 1 \\ 0 \\ 0 \end{pmatrix} \quad \text{and} \quad \begin{pmatrix} 0 \\ -3 \\ 0 \\ 1 \\ 0 \end{pmatrix} \quad \text{and} \quad \begin{pmatrix} -4 \\ -6 \\ 0 \\ 0 \\ 1 \end{pmatrix}$$

The set of all linear combinations (using all arbitrary scalars x_3, x_4, x_5), is called the *span* of the vectors. Notice that each vector in the null space of this matrix is an element of $\mathbb{R}^5$, and $\mathbb{R}^5$ is one of the vector spaces mentioned in the previous section. Our null space is a *subset* of a vector space – the subset consisting of *all* linear combinations of some given vectors. It is not hard to see that every null space can be written in the same way as our example. We will work at least one other example in class.

This pattern occurs in countless examples. In class we will discuss the following sets to reinforce the idea of the span of a set of vectors.

(a) The set of polynomials of degree at most 3.

(b) The set of 3×2 matrices with row 2 all 0's.

(c) The set of vectors in 3-space perpendicular to $3\overrightarrow{i} - \overrightarrow{j}$.

(d) The set of functions $f(t)$ such that $f''(t) = 4 \cdot f(t)$.

(e) The set of all 3×1 matrices B such that the following system is consistent:

$$\begin{pmatrix} 1 & 2 \\ 3 & 4 \\ 5 & 6 \end{pmatrix} \cdot X = B$$

Each of these sets has the following abstract form: it consists of elements of a previously given vector space, and there are particular elements of the set $v_1, v_2, \ldots, v_n$ such that *every* element of the set looks like this:

(8.3) $$a_1 \cdot v_1 + a_2 \cdot v_2 + \cdots + a_n \cdot v_n$$

where $a_1, a_2, \ldots, a_n$ are *arbitrary scalars*. An expression (8.3) is a *linear combination* of $v_1, v_2, \ldots, v_n$. The set of linear combinations is the *span* of these elements.

The main theme of this section: whenever we have the span of given elements in a vector space, the span itself is a vector space. In general, the span will consist of some but not all vectors in the larger space. We say that the span is a *subspace* of the larger vector space.[2]

Let's see that the span of a list of vectors forms a vector space. Suppose we are given $v_1, v_2, \ldots, v_n$ from some vector space V. Let W be the span of $v_1, \ldots, v_n$, so that the elements of W are precisely the linear combinations as in (8.3). The scalars $a_1, \ldots, a_n$ are arbitrary real numbers.

Go back to the paragraph where vector spaces are defined. To be a vector space, the set W needs to have an addition and a scalar multiplication. The elements of W are elements of V, and addition and scalar multiplication are already defined in V; does that suffice? Looking closely at the definition of vector space, if we have given elements of W, their sum *has to be in W*, so that the addition is defined *on W*, not just in V. Look at two elements of W:

$$a_1 \cdot v_1 + \cdots + a_n \cdot v_n \quad \text{and} \quad b_1 \cdot v_1 + \cdots + b_n \cdot v_n$$

---

[2]The words "smaller" and "larger" are being used somewhat informally. The "smaller" vector space has only to be *contained* in the "larger" one. Example: $\mathbb{R}^5$ is a subspace of itself! But in most the cases, the smaller space is only part of the larger one.

Their sum is

$$(a_1 + b_1) \cdot v_1 + (a_2 + b_2) \cdot v_2 + \cdots + (a_n + b_n) \cdot v_n$$

and this sum is yet another linear combination of the same particular vectors: the v_j. In other words, their sum *is* in W. Thus, W has an addition.

Similarly, the scalar multiplication in V is inherited by W, for we can multiply a linear combination by a number r to get

$$r \cdot \big(a_1 \cdot v_1 + \cdots + a_n \cdot v_n\big) = (r \cdot a_1) \cdot v_1 + \cdots + (r \cdot a_n) \cdot v_n$$

and obtain yet another linear combination.

What we have just shown is that the span of a set of vectors has an addition and scalar multiplication defined on it. To be a vector space, these operations need to satisfy properties V1-V7 listed above. Every property except for (V3) holds for all vectors in V and all scalars, and so those properties hold automatically in W. Property V3 says that a vector space must have a zero vector. In our case, the zero vector of V must be an element of W. Observe that

$$\mathbb{O} = 0 \cdot v_1 + 0 \cdot v_2 + \cdots + 0 \cdot v_n$$

so that the zero vector is a linear combination of $v_1, \ldots, v_n$, as needed. Thus, the span of a finite set of vectors in a larger vector space is, itself, a vector space.

Here is an important example: the *row space* of a matrix A is the span of the rows of A.

A vector space that is the span of a given, finite set of vectors is called a *finite dimensional vector space*. We will explain dimension in the next chapter. For now, we want to get used to the ideas of linear combination and span.

It is useful to work the following kind of problem – this sort of problem seems a bit artificial once you understand the subject better.

Problem. Is $x^2 + 1$ in the span of $2x^3 + 3x^2 - x + 3$, $x^3 + x + 1$, $-2x^2 + 3x$?

Solution. This span is a subspace of the vector space of all polynomials. It consists of these:

$$a_1 \cdot (2x^3 + 3x^2 - x + 3) + a_2 \cdot (x^3 + x + 1) + a_3 \cdot (-2x^2 + 3x)$$

where a_1, a_2, a_3 are (unknown) scalars. To ask whether $x^2 + 1$ is in this span is to ask whether the equation

$$(8.4) \quad x^2 + 1 = a_1 \cdot (2x^3 + 3x^2 - x + 3) + a_2 \cdot (x^3 + x + 1) + a_3 \cdot (-2x^2 + 3x)$$

has a solution a_1, a_2, a_3. We rewrite the right side, collecting like powers of x:

$$(2a_1 + a_2) \cdot x^3 + (3a_1 - 2a_3) \cdot x^2 + (-a_1 + a_2 + 3a_3) \cdot x + (3a_1 + a_2)$$

and we can write $x^2 + 1 = 0 \cdot x^3 + 1 \cdot x^2 + 0 \cdot x + 1$. Polynomial equality is equality in each power of x, and so the equation (8.4) is this system of equations:

$$\begin{array}{rcrcrcl}
2a_1 & + & a_2 & & & = & 0 \\
3a_1 & & & - & 2a_3 & = & 1 \\
-a_1 & + & a_2 & + & 3a_3 & = & 0 \\
3a_1 & + & a_2 & & & = & 1
\end{array}$$

Elimination finds the solution $a_1 = 1$, $a_2 = -2$, $a_3 = 1$. In other words, equation (8.4) has this form

$$x^2 + 1 = 1 \cdot (2x^3 + 3x^2 - x + 3) - 2 \cdot (x^3 + x + 1) + 1 \cdot (-2x^2 + 3x)$$

So the answer is, "Yes, $x^2 + 1$ is a linear combination of the three polynomials given." ■

It is no coincidence that the previous problem is solved by a system of linear equations. We will consider at least one more such problem in class, and there are some more at the end of the chapter.

Here is a general example like (e) above, but more universal. Let A be an $m \times n$ matrix. Each column of A is an $m \times 1$ matrix, and so their span would consist of various $m \times 1$ matrices. This span is called the *column space* of A. We will now show that the column space of A can be described in another way: as the set of $m \times 1$ matrices B such that $A \cdot X = B$ is consistent. This fact depends on a formula for multiplication. Write the columns of A like this:

$$A = \begin{pmatrix} A_1 & A_2 & \cdots & A_n \end{pmatrix}$$

Given scalars $c_1, \ldots, c_n$, we claim that

(8.5) $$A \cdot \begin{pmatrix} c_1 \\ c_2 \\ \vdots \\ c_n \end{pmatrix} = c_1 \cdot A_1 + c_2 \cdot A_2 + \cdots + c_n \cdot A_n$$

Indeed, the product indicated is $m \times 1$. Its $[i, 1]$-entry is the dot product of row i of A with the column of the c_k's. Notice that $A[i, j] = A_j[i, 1]$, since A_j is the j-th column of A, and so the $[i, 1]$-entry of the left side of (8.5) is this:

$$\sum_{j=1}^{n} A[i, j] \cdot c_j = \sum_{j=1}^{n} A_j[i, 1] \cdot c_j$$

The rightmost sum is the $[i, 1]$ entry of the sum on the right of (8.5), as needed. This formula is related to Proposition 3.1.

Now we can prove that the column space of A is the set of B where $A \cdot X = B$ is consistent. If $AX = B$ is consistent, then there is a solution C with $AC = B$. Equation (8.5) shows that $B = AC$ is a linear combination of the columns of A, since the c_k's are the scalars. Conversely, assume that B is a linear combination of the columns of A. Then (8.5) shows that $A \cdot X = B$ has solution $X_j = c_j$ for each j.

The columns of A are the rows of A^T. Thus, the span of the columns of A is the span of the rows of A^T. (Of course, the columns are vertical and the rows horizontal, but they correspond.) Thus, the column space of A can be identified with the row space of A^T, and the row space of A with the column space of A^T.

It might help to note an example of a vector space that is *not* spanned by a finite set of particular vectors. The set of *all* polynomials is such a space. A finite set of polynomials has a maximum degree k; the polynomial x^{k+1} cannot be a linear combination of polynomials of degree less than or equal to k. Thus, no finite set of polynomials spans the set of all polynomials.

2.1. Abstract Subspaces. Given vectors $v_1, \ldots, v_n$ in a vector space V, we showed that the span of the v_j forms a *subspace* – a subset that forms a vector space in its own right. The concept of subspace is itself interesting, apart from the idea of a span, and so we isolate that idea. Actually, we have done that implicitly when we proved that the span is a subspace. Since not all subspaces come about as the span of a finite set of vectors, the idea of a subspace is more general than the idea of the span.

PROPOSITION 8.1. *Let W be a subset of the vector space V. Then W is a subspace of V if and only if each of the following statements is true.*

  (1) if u, v are in W, then $u + v$ is in W
  (2) if u is in W and if α is a number, then $\alpha \cdot u$ is in W
  (3) $\mathbb{O}$ is in W

PROOF. Suppose that W is a subspace of V. (Most of the following simply rehearses the definition of *subspace*.) The vector addition of V is an operation on W. In other words, if u, v are in W, then $u + v$ is in W, so that (1) holds. Similarly, scalar multiplication is an operation on W, so (2) holds. A vector space has to have a zero vector; $\mathbb{O}$ is the unique zero vector in V, and so it must be in W, and we see that (3) holds.

Now suppose that (1), (2), and (3) hold. Statements (1) and (2) show that the vector addition and scalar multiplication on V are operations on W. Vector space properties V1-V7, except for (V3), hold on W, since they hold on V. As for vector space property (V3), statement (3) above shows that W has the zero vector. $\square$

3. Problems

8.1. Show that the set of differentiable functions on $[0, 1]$ is a vector space. (Describe the addition and scalar multiplication and quote a calculus book to show that they obey the properties V1-V7 given for vector spaces. This is basically a reference exercise!)

8.2. (A crazy example that illustrates the vector space properties.) Let P be the set of positive numbers. For a, b in P, define $a \oplus b = a \cdot b$; for a in P and a real number r, define $r \otimes a = a^r$. Show that P is a vector space using $\oplus$ as addition and $\otimes$ as scalar multiplication. (Note: this is an exercise in very carefully writing the properties (V1)-(V7) for a vector space and translating those properties to real number multiplication and exponentiation.)

8.3. In each case, show that the given set is a finite dimensional vector space: that it is the set of vectors spanned by a finite set of particular instances.
a) The set of polynomials of degree at most 4 having 3 as a root.
b) The set of 3×2 matrices A such that

$$\begin{pmatrix} 1 & 2 & -1 \\ 4 & 8 & -4 \end{pmatrix} \cdot A = \mathbb{O}_{2 \times 2}$$

c) The set of functions $f(t)$ such that $f''(t) + 5 \cdot f'(t) - 14 \cdot f(t) = 0$.

8.4. In each case, determine whether the first vector given is or is not a linear combination of the remaining vectors in the list.
a) $(-1, 1, 3)$, $(1, 2, 3)$, $(4, 5, 6)$, $(7, 8, 9)$.
b) $x^2 + 5$, $x^2 - x$, $x^2 + 2 \cdot x$, $x^2 - x + 1$.
c) $\cos(2x)$, 1, $\sin^2(x)$.

8.5. Show that *every* 3×1 matrix is a linear combination of the columns of the matrix

$$\begin{pmatrix} 1 & 1 & -3 \\ 0 & -1 & 4 \\ 3 & 2 & -3 \end{pmatrix}$$

8.6. Let

$$A = \begin{pmatrix} 2 & 4 & 8 \\ 3 & 1 & 2 \end{pmatrix}$$

Which columns of A can be written as a linear combination of the others? (In each case, show how the column is a linear combination of the others.)

8.7. Let B be a non-zero $m \times 1$ matrix, and let A be an $m \times n$ matrix. Show that the set of solutions to the system $AX = B$ is **not** a vector space.

8.8. Let V be the (fundamental) vector space of differentiable functions defined on the real numbers, and let W be the set of functions $f(t)$ in V such that $f''(t)$ exists. Show that W is a subspace of V.

8.9. Let V be the vector space of all polynomials in the variable t. Let W be the set of polynomials $f(t)$ such that

$$\int_0^1 f(t) \cdot dt = 0$$

Show that W is a subspace of V.

8.10. Let v be a vector in $\mathbb{R}^n$. Define W to be the set of vectors w in $\mathbb{R}^n$ such that $w \circ v = 0$. Show that W is a subspace of $\mathbb{R}^n$.

8.11. Let $v_1, \ldots, v_n$ be vectors in $\mathbb{R}^m$. Let w be in $\mathbb{R}^m$ and suppose that $w \circ v_j = 0$ for each j. Show that $w \circ v = 0$ for every v in the span of $v_1, \ldots, v_n$. Show that the set of all such w forms a vector space.

8.12. Let $v_1, \ldots, v_m$ be the rows of the $m \times n$ matrix A. Let $w \in \mathbb{R}^n$. Show that $w \circ v_j = 0$ for all j if and only if w is in the null space of A.

CHAPTER 9

Dimension and Basis

1. Linear Independence

When we recognize that a set of vectors is the span of a finite set of its vectors, we know we have a vector space. Every vector in the space is a linear combination of the spanning vectors. The *dimension* of a vector space is the *minimum* number of vectors necessary to span it. It will help to begin with an example that shows how a minimum might be arrived at.

Suppose that the vector space V is spanned by vectors v_1, v_2, v_2, v_4. We show that if one of the v_j is a linear combination of the others, then they are **not** a minimum spanning set. For instance, suppose that

$$(9.1) \qquad v_4 = 3 \cdot v_1 - 2 \cdot v_2 + 4 \cdot v_3$$

(We'll see that the specific numbers $3, 2, 4$ can be any numbers.) We can show that v_1, v_2, v_3 span the entire vector space V. In other words, we don't need v_4. To see this, take an arbitrary vector in V:

$$v = a_1 \cdot v_1 + a_2 \cdot v_2 + a_3 \cdot v_3 + a_4 \cdot v_4$$

for numbers a_1, a_2, a_3, a_4. We can use equation (9.1):

$$v = a_1 \cdot v_1 + a_2 \cdot v_2 + a_3 \cdot v_3 + a_4 \cdot (3 \cdot v_1 - 2 \cdot v_2 + 4 \cdot v_3)$$
$$= (a_1 + 3) \cdot v_1 + (a_2 - 2) \cdot v_2 + (a_3 + 4) \cdot v_3$$

We see that every element of V is a linear combination of v_1, v_2, v_3.

It is not hard to see that this works in general.

PROPOSITION 9.1. *Let $v_1, \ldots, v_n$ be vectors in some vector space, and assume that v_n is a linear combination of $v_1, \ldots, v_{n-1}$. Then the span of $v_1, \ldots, v_{n-1}$ is exactly the same as the span of $v_1, \ldots, v_n$.*

If we are given a finite set of vectors from a vector space, we can test them, one by one, to see whether one of them is a linear combination of the others. If one of them is such a combination, Proposition 9.1 shows that it can be dropped from the list without changing the span. You might see how this makes the span more *efficient* – fewer vectors are needed to express the elements of the span. We can apparently keep dropping elements from the list of spanning vectors until it is no longer possible – none of the vectors is a linear combination of the others. It turns out that this list of vectors is minimal – it counts the dimension of the vector space spanned. To see this, we need to introduce another concept.

Let $v_1, v_2, \ldots, v_n$ be vectors from a vector space. We say that these vectors are *linearly independent* if the only way to write the zero vector as a linear combination of $v_1, \ldots, v_n$ is to use all 0 scalars. In other words, the equation

$$(9.2) \qquad \mathbb{O} = a_1 \cdot v_1 + a_2 \cdot v_2 + \cdots + a_n \cdot v_n$$

has only one solution: $0 = a_1 = a_2 = \cdots = a_n$. The direction of the logic here is important: we are **not** saying that $0 = a_1 = a_2 = \cdots$ is a solution to (9.2) – that fact is true no matter what the v_j are! We are saying that the v_j are *linearly independent* when the *only solution* to (9.2) is to have all the a_j equal to 0.

Problem. Show that $x^3 + x$, $x^3 - 2x^2$, x^2 are linearly independent.

Solution. Equation (9.2) looks like this:

$$0 = a_1 \cdot (x^3 + x) + a_2 \cdot (x^3 - 2x^2) + a_3(x^2)$$

(The 0 on the left is the 0 polynomial: the polynomial all of whose coefficients are 0.) We collect the right side terms by powers of x:

$$0 = (a_1 + a_2) \cdot x^3 + (-2a_2 + a_3) \cdot x^2 + a_1 \cdot x$$

The coefficients must be 0, and we get these equations.

$$\begin{bmatrix} 1 & 1 & 0 \\ 0 & -2 & 1 \\ 1 & 0 & 0 \end{bmatrix} \cdot \begin{bmatrix} a_1 \\ a_2 \\ a_3 \end{bmatrix} = \mathbb{0}_{3 \times 1}$$

Elimination shows that the only solution is $a_1 = a_2 = a_3 = 0$, and that proves that the polynomials are independent. ∎

Problem. Are these elements of $\mathbb{R}^2$ linearly independent? $(1, 2)$, $(-2, 1)$, $(3, 4)$

Solution. We should look at the equation

$$(0, 0) = a_1 \cdot (1, 2) + a_2 \cdot (-2, 1) + a_3 \cdot (3, 4)$$

If this equation has **one** solution ($a_1 = a_2 = a_3 = 0$) then the vectors are linearly independent. If the equation has more than one solution, the vectors are dependent. The equation:

$$\begin{pmatrix} 1 & -2 & 3 \\ 2 & 1 & 4 \end{pmatrix} \cdot \begin{pmatrix} a_1 \\ a_2 \\ a_3 \end{pmatrix} = \begin{pmatrix} 0 \\ 0 \end{pmatrix}$$

Elimination shows that a_3 is free, and so there are infinitely many solutions. The vectors are **not** independent.[1] ∎

Here is the relationship between being linearly independent and being a minimal spanning set.

---

[1]When a list of vectors is not independent, it can be said to be *dependent*. We will not use this term, preferring to emphasize the definition of linear independence.

PROPOSITION 9.2. *Let $v_1, v_2, \ldots, v_n$ be vectors from a vector space. These vectors are linearly independent if and only if no one of them is a linear combination of the others.*

PROOF. Suppose first that they are linearly independent, and imagine trying to show that v_n, say, is a linear combination of the others. We would try to solve the equation

$$v_n = a_1 \cdot v_1 + a_2 \cdot v_2 + \cdots + a_{n-1} \cdot v_{n-1}$$

We can write this equation

$$\mathbb{O} = a_1 \cdot v_1 + a_2 \cdot v_2 + \cdots + a_{n-1} \cdot v_{n-1} + (-1) \cdot v_n$$

Since the v_j are linearly independent, there is no linear combination of them that gives the zero vector $\mathbb{O}$ except when all the scalars are 0. The equation we just wrote uses -1 as scalar on v_n, and this is a contradiction. Thus, v_n is not a linear combination of the other vectors. Similarly, no one of the vectors is a linear combination of the others.

Now assume that no one vector of the v_j is a linear combination of the others, and we will show that they are linearly independent. To do that, we need to consider the equation

$$\mathbb{O} = a_1 \cdot v_1 + a_2 \cdot v_2 + \cdots + a_n \cdot v_n$$

We want to show that all the a_j must be 0. Indeed, suppose, for instance, that there is a solution in which a_1 is not 0. Then we can move $a_1 \cdot v_1$ to the left side and divide by $-a_1$:

$$-a_1 \cdot v_1 = a_2 \cdot v_2 + \cdots + a_n \cdot v_n$$

$$v_1 = -(a_2/a_1) \cdot v_2 + \cdots - (a_n/a_1) \cdot v_n$$

This shows that v_1 is a linear combination of the other vectors – and we know that's not the case. Thus, a_1 must be 0, and so must all the scalars be. The vectors $v_1, \ldots, v_n$ are linearly independent.   □

Here is what we have done so far. Given vectors $v_1, \ldots, v_n$, we form the vector space V from their span. If one of the v_j is a linear combination of the others, it can be deleted without changing the span V. We can keep doing this until we arrive at a set of vectors such that no one of them is a linear combination of the others. Proposition 9.2 says that these vectors are linearly independent. They also span V. A set of vectors that is linearly independent and spans a vector space V is said to be a *basis* for V. A basis counts the *dimension* of the vector space, as we will see soon.

2. Vector Space Basis

First, we give an example to show that a vector space can have many different bases. It is not hard to show that

$$\begin{pmatrix} 1 \\ 0 \\ 0 \end{pmatrix}, \quad \begin{pmatrix} 0 \\ 1 \\ 0 \end{pmatrix}, \quad \begin{pmatrix} 0 \\ 0 \\ 1 \end{pmatrix}$$

form a basis for $\mathbb{R}^3$. (Check this! Do they *span* $\mathbb{R}^3$? Are they linearly independent? Just use the definitions of these terms.) Here is *another* example of a basis for $\mathbb{R}^3$

$$\begin{pmatrix} 2 \\ 1 \\ 0 \end{pmatrix}, \quad \begin{pmatrix} -1 \\ 1 \\ 0 \end{pmatrix}, \quad \begin{pmatrix} 1 \\ -1 \\ 1 \end{pmatrix}$$

Again, showing that this is a basis is straightforward: we show that it spans and that the vectors are linearly independent. So $\mathbb{R}^3$ does not have a *unique* basis. But notice that both bases have 3 vectors in them. The dimension of $\mathbb{R}^3$ is 3 (not a surprise), and that's why there have to be 3.

Here is the general result. It is sometimes called the *Fundamental Theorem of Linear Algebra*; a title that may seem to overstate things a bit.

PROPOSITION 9.3. *Every finite dimensional vector space has a basis, and every basis for the vector space has the same number of vectors. The dimension of the vector space is the number of vectors in each basis.*

To prove Proposition 9.3, we will need the following – it will be used in its own right later. We think of it as saying that spanning sets are *large* and independent sets are *small*.

PROPOSITION 9.4. *Let $w_1, \ldots, w_m$ span the vector space V, and let the vectors $v_1, \ldots, v_n$ in V be linearly independent. Then $n \leq m$.*

PROOF. Each v_j is a linear combination of the w_i's. Find scalars $A[i,j]$ such that

$$(9.3) \qquad v_j = \sum_{i=1}^{m} A[i,j] \cdot w_i \quad \text{for} \quad 1 \leq j \leq n$$

Keep your eye on the matrix A, which is about to become the coefficient matrix of a system of equations. Notice that A is $m \times n$. Assume that C is $n \times 1$ and that $A \cdot C = \mathbb{O}_{m \times 1}$. We will show that C is zero! The key to this is to use the entries of C as coefficients on the v_j. Here goes, using the equations in (9.3) to switch from the v_j's to the w_i's, and then using $A \cdot C = \mathbb{O}$.

$$\sum_{j=1}^{n} C[j,1] \cdot v_j = \sum_{j=1}^{n} C[j,1] \cdot \left(\sum_{i=1}^{m} A[i,j] \cdot w_i \right)$$

$$= \sum_{j=1}^{n} \sum_{i=1}^{m} C[j,1] \cdot A[i,j] \cdot w_i$$

$$= \sum_{i=1}^{m} \sum_{j=1}^{n} A[i,j] \cdot C[j,1] \cdot w_i$$

where the summations were switched to get the last expression.

Continuing:

$$= \sum_{i=1}^{m} \left(\sum_{j=1}^{n} A[i,j] \cdot C[j,1] \right) \cdot w_i$$

$$= \sum_{i=1}^{m} (0) \cdot w_i = \mathbb{O}$$

We conclude that

$$\sum_{j=1}^{n} C[j,1] \cdot v_j = \mathbb{O}$$

and since the v_j's are independent, it must be that all the $C[j,1]$ are 0. In other words, $C = \mathbb{O}$.

We showed that if $A \cdot C = \mathbb{O}$, then $C = \mathbb{O}$. In other words, the system of equations $A \cdot X = \mathbb{O}$ has exactly one solution. As we have seen before, this says that A has rank n. Since A has m rows, it must be that $n \leq m$, as needed. □

Now we can prove Proposition 9.3.

Proof of Proposition 9.3   Let V be a vector space spanned by some finite set of vectors. We defined the dimension as the minimum number of vectors needed to span the vector space: let $v_1, \ldots, v_n$ be as few vectors as possible to span V, so that the dimension of V is n. We claim that these vectors form a basis. They span V; are they linearly independent? If not, then Proposition 9.2 shows that one of them is a linear combination of the others. Proposition 9.1 then says we can delete this vector and have the same span. This contradicts that n is the minimum number of vectors that spans V. Thus, these vectors are linearly independent, and so they form a basis.

Let $w_1, \ldots, w_m$ be a basis for V and we will show that $n = m$. We note that it is *not necessarily true* that the w_i are themselves the same as the v_j. We will address this by examples shortly. For now we prove that $m = n$. Indeed, the $w_1, \ldots, w_m$ span V and $v_1, \ldots, v_n$ are independent. Proposition 9.4 says that $n \leq m$. Reversing roles: $v_1, \ldots, v_n$ span and $w_1, \ldots, w_m$ are independent, so $m \leq n$. ∎

Problem. Show that $(1, 2)$, $(3, 4)$ is a basis for $\mathbb{R}^2$.

Solution. We will work directly from the definition: the vectors need to span $\mathbb{R}^2$ and be linearly independent. Independence first: we consider this equation

$$(0, 0) = a_1 \cdot (1, 2) + a_2 \cdot (3, 4) = (a_1 + 3a_2, 2a_1 + 4a_2)$$

solving for a_1, a_2. The resulting system of equations:

$$\begin{pmatrix} 1 & 3 \\ 2 & 4 \end{pmatrix} \cdot \begin{pmatrix} a_1 \\ a_2 \end{pmatrix} = \begin{pmatrix} 0 \\ 0 \end{pmatrix}$$

has rank 2 and a unique solution $a_1 = 0 = a_2$. Uniqueness here shows that the given pair of vectors are linearly independent. Do the vectors span $\mathbb{R}^2$? Given an arbitrary element (b, c) of $\mathbb{R}^2$, we need to find scalars a_1, a_2 such that

$$(b, c) = a_1 \cdot (1, 2) + a_2 \cdot (3, 4) = (a_1 + 3a_2, 2a_1 + 4a_2)$$

The same system of equations with an arbitrary right side! This time, the fact that the rank is 2 shows that we get a solution a_1, a_2 for every b, c. The vectors span $\mathbb{R}^2$ and so they give a basis for $\mathbb{R}^2$. ∎

Proposition 9.3 is a very important fact – it is sometimes called *The Fundamental Theorem of Vector Spaces*, and it has been more humorously referred to as *the only theorem of vector spaces*. The only trouble is that it doesn't tell us how to find a basis. Fortunately, Elimination handles a particular case that generalizes to many situations. The particular case involves the basis for the row space of a matrix.

PROPOSITION 9.5. *Let A be a non-zero $m \times n$ matrix. Perform Elimination on A to get a row-echelon form A'. Then the non-zero rows of A' form a basis for the row space of A. In particular, the dimension of the row space of A is the rank of A.*

PROOF. Suppose we perform an elementary operation on A to get the matrix B. We claim that the row space of A is exactly that of B. This is obvious in the case of switching rows, since that doesn't change the vectors spanning the space. It is also obvious in the case of non-zero multiplication of one row by a scalar.

To handle the case of adding a multiple of one row to another, denote row j of A by A_j, and enforce a similar notation on B. Say that row 1 of A is replaced by $A_1 + \beta \cdot A_2$, for some number β. In other words, we are assuming that $B_1 = A_1 + \beta \cdot A_2$. And $B_j = A_j$ for $j = 2, 3, \ldots$. It is easy to see that linear combinations of the B_j are linear combinations of the A_j, and so the row space of B is contained in the row space of A. We know that A can be obtained from B by an inverse elementary operation, and so the row space of A is contained in the row space of B. This shows that the two row spaces are equal.

Applying the fact of the previous paragraph over and over, we see that A and its row-echelon form A' have the same row space. Thus, a basis for the row space of A' will be a basis for the row space of A. In taking the span of the rows of A', we can obviously discard the zero rows. We have only to show that the non-zero rows are linearly independent. Suppose that A' has r non-zero rows (suppose that the rank of A is r), and consider the equation

(9.4) $$\mathbb{O} = c_1 \cdot A'_1 + c_2 \cdot A'_2 + \cdots + c_r \cdot A'_r$$

Consider one of the these rows A'_j. Each non-zero row in row-echelon form has a pivot; say the pivot for A'_j lies in column k. Then $A'[j, k] = 1$ and there are no other non-zero entries in column k. Looking at the k-th entry in (9.4), we see that

$$0 = c_j \cdot A'_j[k] = c_j$$

Each of the c_j is 0, and this shows that the non-zero rows of A' are linearly independent. $\square$

Proposition 9.5 shows that the rank of a matrix does not depend on choices made in Elimination, since the rank is the dimension of a vector space determined by the matrix itself, without respect to any elementary operations.

Here is an interesting application of Proposition 9.5

Problem.   Find a basis for the vector space spanned by the polynomials

$$2 + 3x, \quad 4 - 5x + x^2, \quad x - x^2, \quad 1 + x^2$$

Solution.   The vectors are polynomials of degree at most 2, and we can think of each of them as a 1×3 matrix:

$$
\begin{array}{c}
2 + 3x \\
4 - 5x + x^2 \\
x - x^2 \\
1 + x^2
\end{array}
\quad \text{as} \quad
\begin{bmatrix}
2 & 3 & 0 \\
4 & -5 & 1 \\
0 & 1 & -1 \\
1 & 0 & 1
\end{bmatrix}
$$

In other words, our vector space of polynomials is like the row space of the matrix on the right. Performing Elimination,[2] along the lines of Proposition 9.5, we get row-echelon form

$$
\begin{bmatrix}
1 & 0 & 0 \\
0 & 1 & 0 \\
0 & 0 & 1 \\
0 & 0 & 0
\end{bmatrix}
$$

Turning the rows back into polynomials we have basis 1, x, x^2. ∎

The previous problem is highly suggestive! To find a basis for a vector space, write the vectors as matrix rows and use Elimination. We will formalize this idea in the next chapter.

Here is a more general problem: invertible matrices give bases (the plural of basis) for $\mathbb{R}^n$.

---

[2]We need to see that the row operations correspond to manipulating vectors in our vector space. This is straightforward and will be discussed in class.

Problem. Let A be an $n \times n$ invertible matrix. Show that the columns of A form a basis for $\mathbb{R}^n$.

Solution. The column space of A is the row space of A^T. Proposition 9.5 says that a row-echelon from for A^T will give a basis for its row space – a basis for the column space of A.

Since A inverts, the matrix A^T has an inverse: $(A^{-1})^T$. Thus, A^T has rank n, and its row-echelon form is I_n. The rows of I_n are a basis for $\mathbb{R}^n$, and we see that the row space of A^T is $\mathbb{R}^n$. It follows that the column space of A is $\mathbb{R}^n$. The proof of Proposition 9.3 shows that we can find a basis among the columns of A, we see that those n columns form a basis. ∎

A consequence: there may be many different bases to a given vector space, since there are many different invertible matrices.

The previous paragraph contains a very practical algorithm, for it shows how to find a basis for the column space of a matrix A. The column space has dimension equal to the rank of A^T. Statement (4) says that the dimension of the column space is the rank of A. In other words, (4) claims that the rank of A is that of A^T. This result has its own name.

THE RANK THEOREM. *Let A be an $m \times n$ matrix. Then the dimension of the row space of A is equal to the dimension of its column space. Also, the rank of A is equal to the rank of A^T.*

PROOF. Proposition 9.5 showed that the rank of A is the dimension of its row space. We will show that the rank is the dimension of the column space. Do Elimination on A to produce row-echelon form A', in which columns $c_1, \ldots, c_r$ get pivots. (So the rank of A is r.)

Let C be the matrix of columns $c_1, \ldots, c_r$ of A. Thus, C is $m \times r$. We claim that the column spaces of A and C are exactly the same. If B is in the column space of C, then B is a linear combination of columns of C. The columns of C

are columns of A, and so B is in the column space of A. Conversely, if B is in the column space of A, then Elimination on $A \cdot X = B$ produces $A' \cdot X = B'$, where B' is 0 below row r, since the first r rows of A' are where the pivots ended up. This shows that $C \cdot X = B$ is also consistent, for the row-echelon form of C involves columns $c_1, \dots, c_r$ of A'. Thus, B is in the column space of C, and the two column spaces are the same.

The placement of pivots in the columns of C shows that its r columns are linearly independent. Thus, the dimension of the column space of C is r, and so the dimension of the column space of A is r, as well, as claimed.

The fact about A^T follows, for the rank of A is the dimension of the column space of A, and that is the dimension of the row space of A^T. By Proposition 9.5, the dimension of the row space of A^T is the rank of A^T.   □

Let's review some important cases of dimension. All have been proved except for (2), which will be left to you or to class.

1) The dimension of $\mathbb{R}^n$ is n.

2) The dimension of the space of polynomials of degree at most n is $n + 1$.

3) The dimension of the null space of a matrix is the nullity of the matrix.

4) The dimension of the column space of a matrix is the rank of the matrix.

There is at least one more fact that is sometimes useful. A set of linearly independent vectors can *grow into* a basis.

PROPOSITION 9.6. *Let V be a finite dimensional vector space, and let $v_1, \dots, v_m$ be linear independent elements of V. Then there is a basis of V that includes these vectors.*

PROOF. Let n be the dimension of V. Since $v_1, \dots, v_m$ are independent and since the basis is a spanning set of n vectors, Proposition 9.4 shows that $m \leq n$.

If $v_1, \ldots, v_m$ span V, then they are a basis, and we are done. Otherwise, there must be some vector v in V that is not in the span of $v_1, \ldots, v_m$. We claim that $v, v_1, \ldots, v_m$ are independent. Indeed, if not then we can write

$$\mathbb{O} = c \cdot v + b_1 \cdot v_1 + \cdots b_m \cdot v_m$$

for some scalars $c, b_1, \ldots, b_m$ that are not all 0. If $c = 0$, then one of the b_k is not 0, and this contradicts that $v_1, \ldots, v_m$ are independent. Thus, $c \neq 0$, and, repeating a manipulation we gave before,

$$v = -\frac{1}{c} \cdot \left(b_1 \cdot v_1 + \cdots b_m \cdot v_m \right)$$

and this shows that v is in the span of $v_1, \ldots, v_m$. This contradiction shows that $v, v_1, v_2, \ldots, v_m$ are independent.

If the resulting $m + 1$ vectors span V, we have the desired basis. Otherwise, find another vector not in the span, and include it to get a larger set of independent vectors. This process cannot be repeated indefinitely, since Proposition 9.4 shows that n is the upper limit on the number of independent vectors. The process must therefore terminate, and when it does, we have a basis that includes the original independent vectors. □

3. Problems

9.1. Which of the following are linearly independent?

(a) $(0, 1, 3), (1, 3, 7), (5, 0, 1)$
(b) the polynomials $X^2, X^2 + 3, X^2 - 3$
(c) the functions $\cos(x), \sin(x)$
(d) the columns of an non-invertible 4×4 matrix

9.2. Find a basis for the vector space spanned by these matrices

$$\begin{pmatrix} 1 & 2 & 3 \\ 4 & 5 & 6 \end{pmatrix}, \quad \begin{pmatrix} -1 & -2 & 3 \\ 1 & 2 & 1 \end{pmatrix}, \quad \begin{pmatrix} -1 & -2 & 9 \\ 6 & 9 & 8 \end{pmatrix}$$

9.3.   Find a basis for the null space of the following matrix:

$$\begin{pmatrix} 1 & 1 & 2 & 4 \\ 1 & -2 & 1 & 3 \\ -5 & 8 & 1 & -1 \end{pmatrix}$$

9.4.   Find a basis for the vector space of solutions to this DE:

$$y''' + 6 \cdot y'' + 8 \cdot y' = 0$$

9.5.   Find a basis for the column space of this matrix.

$$\begin{pmatrix} 1 & 7 & -5 & -1 \\ -4 & 4 & 3 & -1 \\ 23 & 1 & 0 & 2 \end{pmatrix}$$

9.6.   Find a basis for the vector space in $\mathbb{R}^3$ consisting of vectors perpendicular to $(-3, 1, 2)$.

9.7.   Find a basis for the vector space of polynomials of degree at most 4 that have -3 as a root.

9.8.   Let $v_1, \ldots, v_n$ be linearly independent vectors in some vector space. Suppose there are scalars a_j and b_j such that

$$a_1 \cdot v_1 + \cdots + a_n \cdot v_n = b_1 \cdot v_1 + \cdots + b_n \cdot v_n$$

Show that $a_1 = b_1$ and $a_2 = b_2$, and so on. (Hint: bring all terms to the left side.)

9.9.   Suppose that v_1, v_2, v_3 is a basis for the vector space V. Show that

$$v_1 + v_2, \quad v_2 + v_3, \quad v_3$$

is also a basis for V. (Hint: make direct use of both aspects of the definition of *basis*.)

9.10. Follow the steps given to prove the following fact: let V be a subspace of $\mathbb{R}^n$ and define W to be the set of w in $\mathbb{R}^n$ such that $w \circ v = 0$ for all $v \in V$. We can show that W is a vector space; assume that for now. Let p be the dimension of V, and let q be the dimension of W. Then $p + q = n$.

(a) Let $v_1, \ldots, v_p$ be a basis for V. Write each v_j as $1 \times n$ and form them into the rows of a $p \times n$ matrix A. Then V is the row space of A.

(b) The rank of A is p. (Hint: the rank theorem.)

(c) The set W is the null space of A.

(d) The dimension of W is the nullity of A.

(e) We have $p + q = n$.

CHAPTER 10

Linear Transformations

We begin with the abstract notation for a function. If we have sets A, B, then the notation $f : A \to B$ means that for each element a of A, there is an element $f(a)$ of B. We say that f is a *function* from A to B. Another matter: we will continue to think of the elements of $\mathbb{R}^n$ as $n \times 1$ matrices.

1. Examples and the Definition

We introduced vector spaces by noticing that several different sets have an *addition* and *scalar multiplication*; the common features of these operations were abstracted to produce the idea of a *vector space*. We will arrive at the idea of a *linear transformation* in a similar way. As with vector spaces, so linear transformations are ubiquitous, for they are the natural functions from one vector space to another. Consider the following functions:

(a) Let P_4 be the vector space of polynomials of degree at most 4, and let P_3 be the polynomials of degree at most 3. Define $L : P_4 \to P_3$ by letting $L(g(t)) = g'(t)$ for each $g(t)$ in P.

(b) Let A be an $m \times n$ matrix, and define $f : \mathbb{R}^n \to \mathbb{R}^m$ by $f(v) = A \cdot v$.

(c) Define $r : \mathbb{R}^2 \to \mathbb{R}^2$ by letting $r(u)$ be the reflection of u across the line $y = 3 \cdot x$.

For example (c), we need to remember some vector algebra. Let $u, v \in \mathbb{R}^2$ with $v \neq \mathbb{O}$, and recall the *projection of u onto v*:

$$w = \frac{u \circ v}{v \circ v} \cdot v$$

The vector w is in the same direction as v, or in the opposite direction, and the vectors v and $w-u$ are perpendicular, as is shown by calculating $v \circ (w-u) = 0$.

The reflection of u by the line of scalar multiples of v is easy to picture; we will do so in class. We will see that

$$r(u) = u + 2 \cdot (w - u) = 2 \cdot w - u = 2 \cdot \frac{u \circ v}{v \circ v} \cdot v - u$$

so that

(10.1) $$r(u) = 2 \cdot \frac{u \circ v}{v \circ v} \cdot v - u$$

In (c), the line is $y = 3 \cdot x$, and we can take $v = (1,3)$.

The sets involved in the three examples: P_4, P_3, and the various $\mathbb{R}^n$, are all vector spaces. In each case, we have a function $f : U \to V$ where U, V are vector spaces, and we have the following properties.

Property 1. $f(a + b) = f(a) + f(b)$ for all a, b in U.

Property 2. $f(\beta \cdot a) = \beta \cdot f(a)$ for all a in U and all real numbers β.

It is a direct calculation to show that our three examples have Property 1 and Property 2. Indeed, for example (a), Property 1 says that

$$\big[g(t) + h(t)\big]' = g'(t) + h'(t)$$

For example (c), we use (10.1) to verify Property 2,

$$r(\beta \cdot u) = 2 \cdot \frac{(\beta \cdot u) \circ v}{v \circ v} \cdot v - (\beta \cdot u) = \beta \cdot \left[2 \cdot \frac{u \circ v}{v \circ v} \cdot v - u\right] = \beta \cdot r(u)$$

When U, V are vector spaces and $f : U \to V$ has Property 1 and Property 2, we say that f is a *linear transformation* from U to V. To get started in understanding these functions, we derive two simple but important additional properties. First additional property:

Claim. Every linear transformation maps the zero vector to the zero vector.

PROOF. If $\mathbb{O}_U$ is the zero vector for U and $\mathbb{O}_V$ is the zero vector for V, and if $f : U \to V$ is a linear transformation, then we claim that

$$f(\mathbb{O}_U) = \mathbb{O}_V$$

To prove this, we use Property 2:

$$f(\mathbb{O}_U) = f(0 \cdot \mathbb{O}_U) = 0 \cdot f(\mathbb{O}_U) = \mathbb{O}_V$$

$\square$

Let's think about the last equality: the vector $f(\mathbb{O}_U)$ is in V (since f maps U to V by definition!). The zero number multiplied by a vector gives the zero vector. Since $f(\mathbb{O}_U)$ is in V, the product gives the zero vector of V.

Our second additional property:

Claim. The composite of linear transformations is a linear transformation.

PROOF. Let $f : U \to V$ and $g : V \to W$ be linear transformations. (Notice the V common to each.) If u is in U, then $f(u)$ is in V, and so $g(f(u))$ is defined as an element of W. Thus, there is a *composite function* $g(f) : U \to W$. We claim that the composite function $g(f(u))$ is a linear transformation. This is not hard; here is Property 1. Let u, v be elements of U. We use Property 1 on f and on g.

$$g(f(u+v)) = g(f(u) + f(v)) = g(f(u)) + g(f(v))$$

Similarly, let β be a real number and observe that

$$g(f(\beta \cdot u)) = g(\beta \cdot f(u)) = \beta \cdot g(f(u))$$

so that Property 2 holds on the composite function $g(f(u))$. $\square$

Later in this chapter we will encounter sequences of three composites! The many parentheses that result will be cumbersome, and so we will do without them. From now on, we will write the value $f(u)$ of a linear transformation as $f \cdot u$. So, the composite $g(f(u))$ will be written $g \cdot f \cdot u$.

Examples We will discuss some of these typical linear transforms in class.

(1) A 3-dimensional rotation $R : \mathbb{R}^3 \to \mathbb{R}^3$.

(2) Projection of a point in $\mathbb{R}^3$ onto a plane through the origin.

(3) The mapping $J : V \to \mathbb{R}$ defined by $J(f(t)) = \int_2 f(t) \cdot dt$, where V is the vector space of continuous functions on the reals.

(4) An operator polynomial $f(D)$ applied to the vector space of infinitely differentiable functions.

We need one more fact; it involves the special case that the linear transformation $f : U \to V$ has an *inverse*. That means that for each v in V, there is one and only one u in U, such that $f \cdot u = v$. In the three examples (a), (b), and (c), at the beginning of this chapter, notice that the reflection in (c) has an inverse. (What is it?) In (b), the mapping $f \cdot v = A \cdot v$ has an inverse if A is invertible.

If $f : U \to V$ has an inverse, and if v is in V, then we write $f^{-1} \cdot v = u$, where u is the unique element of U with $f \cdot u = v$. The functions f and f^{-1} are inverses the way $\exp(x)$ and $\ln(x)$ are inverses; this has nothing to do with numerical inverses. We have

$$f \cdot f^{-1} \cdot v = v \quad \text{for all} \quad v \quad \text{in} \quad V$$

$$f^{-1} \cdot f \cdot u = u \quad \text{for all} \quad u \quad \text{in} \quad U$$

Here is our fact.

PROPOSITION 10.1. *If the linear transformation $f : U \to V$ has an inverse f^{-1}, then f^{-1} is also a linear transformation.*

PROOF. Notice that $f^{-1} : V \to U$. To show that f^{-1} is a linear transformation, we need to verify Properties 1 and Property 2. For Property 1, let

v_1, v_2 be vectors in V, and we need to show that

(10.2) $$f^{-1} \cdot (v_1 + v_2) = f^{-1} \cdot v_1 + f^{-1} \cdot v_2$$

Apply f to the right side and use that f satisfies Property 1:

$$f \cdot (f^{-1} \cdot v_1 + f^{-1} \cdot v_2) = f \cdot f^{-1} \cdot v_1 + f \cdot f^{-1} \cdot v_2 = v_1 + v_2$$

The vector $v_1 + v_2$ is in V, and so there is a unique vector in U which f maps to it. That unique vector is seen to be $f^{-1} \cdot v_1 + f^{-1} \cdot v_2$, but it is also $f^{-1} \cdot (v_1 + v_2)$, by definition. This establishes (10.2).

Next we need to show that if v is in V and β is a real number, then

(10.3) $$f^{-1} \cdot \beta \cdot v = \beta \cdot f^{-1} \cdot v$$

We apply f to the right side and use Property 2 on f.

$$f \cdot \beta \cdot f^{-1} \cdot v = \beta \cdot f \cdot f^{-1} \cdot v = \beta \cdot v$$

But f also sends $f^{-1} \cdot \beta \cdot v$ to $\beta \cdot v$. This proves (10.3). We have proved that the inverse of a linear transformation is a linear transformation. □

2. Matrix Representation

The main idea of this chapter is that *every linear transformation involving finite dimensional vector spaces can be represented by a matrix*. We will need to explain what the word *represented* means; we can give the general idea by saying that questions about linear transformations can (almost always) be changed into questions about the representing matrix. In other words, we can use all we know about matrices (rank, inverses, determinant, etc.) to understand linear transformations.

To begin, we show the prevalence of matrices: *every* linear transformation on the real spaces comes about by matrix multiplication.

PROPOSITION 10.2. *Let $f : \mathbb{R}^n \to \mathbb{R}^m$ be a linear transformation. Then there is an $m \times n$ matrix A such that $f \cdot v = A \cdot v$ for every v in $\mathbb{R}^n$.*

PROOF. Let E_j be the j-th column of the identity matrix I_n, and define $A_j = f \cdot E_j$ for each j. Thus, A_j is an $m \times 1$ matrix, and so we can form an $m \times n$ matrix A using the A_j as the columns. Let v be $n \times 1$, and we use Properties 1 and 2 to compute:

$$f \cdot v = f \cdot \sum_{j=1}^{n} v[j] \cdot E[j] = \sum_{j=1}^{n} f \cdot (v[j] \cdot E[j])$$

$$= \sum_{j=1}^{n} v[j] \cdot f \cdot E[j] = \sum_{j=1}^{n} v[j] \cdot A_j = A \cdot v$$

as needed.   □

Notice that f maps $\mathbb{R}^n$ to $\mathbb{R}^m$ and A comes out $m \times n$. We might remember that the n and m are reversed in the matrix size.

Let's consider Example (c) from the beginning of the chapter. We have $r : \mathbb{R}^2 \to \mathbb{R}^2$ defined by reflection across the line $y = 3 \cdot x$. Using notation from the proof of Proposition 10.2,

$$E_1 = \begin{bmatrix} 1 \\ 0 \end{bmatrix} \quad \text{and} \quad E_2 = \begin{bmatrix} 0 \\ 1 \end{bmatrix}$$

We can use (10.1) with $v = (1, 3)$ to compute that

$$r \cdot E_1 = \begin{bmatrix} -4/5 \\ 3/5 \end{bmatrix} \quad \text{and} \quad r \cdot E_2 = \begin{bmatrix} 3/5 \\ 4/5 \end{bmatrix}$$

The matrix A produced by Proposition 10.2 is

$$A = \begin{bmatrix} -4/5 & 3/5 \\ 3/5 & 4/5 \end{bmatrix}$$

To emphasize the role of this matrix, let's reflect the point $(2, 5)$ across $y = 3 \cdot x$; we have only to multiply:

$$r \cdot \begin{bmatrix} 2 \\ 5 \end{bmatrix} = \begin{bmatrix} -4/5 & 3/5 \\ 3/5 & 4/5 \end{bmatrix} \cdot \begin{bmatrix} 2 \\ 5 \end{bmatrix} = \begin{bmatrix} 7/5 \\ 26/5 \end{bmatrix}$$

We can compare the answer to what we get from formula (10.1).

The equation $f \cdot v = A \cdot v$, for a transformation f, matrix A, and all vectors v, is often written in simpler form: $f = A$. It pays to remember that f and A are not exactly the same thing, since f is a function, and A is a table of numbers. However, they have the same effect on all the vectors. We will write $f = A$, as is customary, but occasionally we will remind you of the distinction.

Next, we begin to relate arbitrary vector spaces to the real spaces; we introduce the *coordinate transformations* – these are linear transformations that make each vector space look like one of the $\mathbb{R}^n$.

Suppose that V is a finite dimensional vector space, so that we know that V has a basis; say $v_1, \ldots, v_n$ is a basis. (Remember that there are many bases!) We construct a linear transformation $L : \mathbb{R}^n \to V$. Given u in $\mathbb{R}^n$, define

$$L(u) = u[1] \cdot v_1 + u[2] \cdot v_2 + \cdots + u[n] \cdot v_n$$

In other words, we use the coordinates of u as scalars to form a linear combination of the v_j. We call L a *coordinate transformation*. In class we will show that L is a linear transformation.

For instance, let P_3 be the set of polynomials of degree at most 3. This vector space has basis $t^3, t^2, t, 1$. Writing each element of $\mathbb{R}^4$ in terms of its coordinates, we have

$$L \begin{pmatrix} a \\ b \\ c \\ d \end{pmatrix} = a \cdot t^3 + b \cdot t^2 + c \cdot t + d$$

This should seem very natural.

Going back to the general case, we claim that L has an inverse. Indeed, each v in V can be written as a linear combination of the basis:

$$v = a_1 \cdot v_1 + \cdots + a_n \cdot v_n$$

where $a_1, \ldots, a_n$ are real numbers. A homework problem showed that there is exactly one such list of scalars. Thus, to get $L(u) = v$, we would need to have

$$u = \begin{pmatrix} a_1 \\ a_2 \\ \vdots \\ a_n \end{pmatrix}$$

In other words,

$$L^{-1}(a_1 \cdot v_1 + \cdots + a_n \cdot v_n) = \begin{pmatrix} a_1 \\ a_2 \\ \vdots \\ a_n \end{pmatrix}$$

Proposition 10.1 shows that L^{-1} is a linear transformation. (It's not too hard to prove that L^{-1} is a linear transformation right from its definition!)

A coordinate transformation depends on the basis used and on the ordering of vectors in that basis. We have included some problems that address these issues.

Now we are ready to define the representing matrix of a linear transformation. This concept is essential, but we warn you that it may seem complicated at first. Let $f : U \to V$ be a linear transformation, where U, V are finite dimensional vector spaces. Let $G : \mathbb{R}^n \to U$ be a coordinate transformation (so that n is the dimension of U), and let $H : \mathbb{R}^m \to V$ be a coordinate transformation of V (and m is the dimension of V). For each v in $\mathbb{R}^n$, we consider

$$H^{-1} \cdot f \cdot G \cdot v$$

Notice that G sends v into U, then f maps $G \cdot v$ from U to V, and then H^{-1} takes us to $\mathbb{R}^m$. Thus, $(H^{-1} \cdot f \cdot G) : \mathbb{R}^n \to \mathbb{R}^m$.

Proposition 10.1 shows that H^{-1} is a linear transformation. The function $H^{-1} \cdot f \cdot G$ is then the composite of three linear transformations, and so it is a linear transformation. Proposition 10.2 then finds an $m \times n$ matrix A such that

$$H^{-1} \cdot f \cdot G \cdot v = A \cdot v \quad \text{for all} \quad v \quad \text{in} \quad \mathbb{R}^n$$

and we abbreviate this to

$$H^{-1} \cdot f \cdot G = A$$

The matrix A is said to *represent* the linear transformation f.

The representing matrix A depends on the basis of U that gives G, and it depends on the basis of V that gives H. If you change either basis, you change A.

Example. Let P_2 be the polynomials of degree at most 2, and let $f : P_2 \to \mathbb{R}$ be evaluation at -2. Thus, $f \cdot g(t) = g(-2)$ for each polynomial $g(t)$ of degree at most 2. We will use the basis $t^2, t, 1$ for P_2 and get the coordinate transformation

$$G \cdot \begin{pmatrix} a \\ b \\ c \end{pmatrix} = a \cdot t^2 + b \cdot t + c$$

For the other vector space $\mathbb{R}$, we use basis 1. The coordinate transformation is the identity:

$$H(a) = a \quad \text{so that} \quad H^{-1}(a) = a$$

Because H and H^{-1} are the identity map, we can omit them; that will simplify our calculation. Now form the composite, remembering that f plugs in $t = -2$.

$$f \cdot G \cdot \begin{pmatrix} a \\ b \\ c \end{pmatrix} = f \cdot (a \cdot t^2 + b \cdot t + c) = 4 \cdot a - 2 \cdot b + c$$

As in the proof of Proposition 10.2, we can use the columns of I_3 to find the 1×3 matrix A that represents f:

$$A = \begin{bmatrix} 4 & -2 & 1 \end{bmatrix}$$

We should see that $f \cdot G = A$.

Here is a typical use of a representing matrix. The question we are going to ask is not profound; the point we are making is that the question can be translated to a matrix question. Let's ask, when do we get $p(-2) = 0$, for

$p(t)$ in P_2? Let's show that G moves this question from f to A. Indeed, if $f \cdot p(t) = p(-2) = 0$, then

$$0 = f \cdot p(t) = f \cdot G \cdot G^{-1} \cdot p(t) = A \cdot G^{-1} \cdot p(t)$$

This says that $G^{-1} \cdot p(t)$ is an element of $\mathbb{R}^3$ in the null space of A.

So, let's find the null space of A. Using the entry 1 as a pivot, we get these solutions.

$$\begin{bmatrix} a_1 \\ a_2 \\ 2 \cdot a_2 - 4 \cdot a_1 \end{bmatrix} \quad \text{with} \quad a_1, a_2 \quad \text{arbitrary}$$

Let v be one of these, and observe that

$$0 = A \cdot v = f \cdot G \cdot v$$

In other words, the polynomial $G \cdot v$ gives 0 when -2 is plugged in. Here is what $G \cdot v$ looks like.

$$a_1 \cdot t^2 + a_2 \cdot t + 2 \cdot a_2 - 4 \cdot a_1 = a_1 \cdot (t^2 - 4) + a_2 \cdot (t + 2)$$

We are not surprised to see these polynomials, for we expect to see factors $t+2$ when -2 is a root. The point is that the matrix A holds this information: we changed the polynomial question, "When is -2 a root?" into the matrix question, "What are the solutions to $A \cdot X = 0$?" ■

Let's do a more sophisticated example.

Example.   Let Q be the vector space of functions $(a + b \cdot t) \cdot e^{3 \cdot t}$, and let $D - 2$ be the operator polynomial, as in Chapter 7. Show that $D - 2$ is a linear transformation mapping Q to Q, and find a representing matrix.

Solution.   We have already referred to operator polynomials as linear transformations. For instance, if $f(t)$ and $g(t)$ are differentiable functions, then

$$(D - 2)(f(t) + g(t)) = f'(t) + g'(t) - 2(f(t) + g(t))$$
$$= f'(t) - 2 \cdot f(t) + g'(t) - 2 \cdot g(t)$$
$$= (D - 2) \cdot f(t) + (D - 2) \cdot g(t)$$

for Property 1. Property 2 is similarly easy. Next, the claim is made that this linear transformation maps Q to Q. In other words, if v is in Q, then $(D-2) \cdot v$ is in Q, as well. We can write $v = (a + b \cdot t) \cdot e^{3 \cdot t}$ for scalars a, b, and compute directly

$$(D - 2) \cdot (a + b \cdot t) \cdot e^{3 \cdot t} = (a + b + b \cdot t) \cdot e^{3 \cdot t}$$

and this last vector is in Q.

To get a representing matrix, we need a basis for Q; we'll use e^{3t}, $t \cdot e^{3t}$. The coordinate transformation G looks like this.

$$G \cdot \begin{bmatrix} a \\ b \end{bmatrix} = (a + b \cdot t) \cdot e^{3 \cdot t}$$

Because we are using the same basis for both instances of Q, we see that $H = G$. Also, observe that

$$G^{-1} \cdot (a + b \cdot t) \cdot e^{3 \cdot t} = \begin{bmatrix} a \\ b \end{bmatrix}$$

To get the matrix that represents $D - 2$, we need to calculate $G^{-1} \cdot (D - 2) \cdot G$ on each column of I_2.

$$G^{-1} \cdot (D - 2) \cdot G \cdot \begin{bmatrix} 1 \\ 0 \end{bmatrix} = G^{-1} \cdot (D - 2) \cdot e^{3t}$$
$$= G^{-1} \cdot \left(3 \cdot e^{3t} - 2 \cdot e^{3t} \right)$$
$$= G^{-1} \cdot e^{3t} = \begin{bmatrix} 1 \\ 0 \end{bmatrix}$$

We also compute that

$$G^{-1} \cdot (D - 2) \cdot G \cdot \begin{bmatrix} 0 \\ 1 \end{bmatrix} = \begin{bmatrix} 1 \\ 1 \end{bmatrix}$$

Therefore, the representing matrix is

$$A = \begin{bmatrix} 1 & 1 \\ 0 & 1 \end{bmatrix}$$

■

Let's give an example use of the matrix. Notice that A has an inverse:

$$A^{-1} = \begin{bmatrix} 1 & -1 \\ 0 & 1 \end{bmatrix}$$

Now consider the non-homogeneous DE

$$y' - 2 \cdot y = 5 \cdot t \cdot e^{3t}$$

We write this $(D - 2) \cdot y = 5 \cdot t \cdot e^{3t}$. We use the matrix A to replace $D - 2$, and then do some algebra using the inverses. Notice that we can solve $G^{-1} \cdot (D - 2) \cdot G = A$ for $D - 2$:

$$D - 2 = G \cdot A \cdot G^{-1}$$

Here's the algebra.

$$(D - 2) \cdot y = 5 \cdot t \cdot e^{3t}$$

$$G \cdot A \cdot G^{-1} \cdot y = 5 \cdot t \cdot e^{3t}$$

$$A \cdot G^{-1} \cdot y = G^{-1} \cdot 5 \cdot t \cdot e^{3t}$$

$$A \cdot G^{-1} \cdot y = \begin{bmatrix} 0 \\ 5 \end{bmatrix}$$

$$G^{-1} \cdot y = A^{-1} \cdot \begin{bmatrix} 0 \\ 5 \end{bmatrix}$$

$$G^{-1} \cdot y = \begin{bmatrix} 1 & -1 \\ 0 & 1 \end{bmatrix} \cdot \begin{bmatrix} 0 \\ 5 \end{bmatrix} = \begin{bmatrix} -5 \\ 5 \end{bmatrix}$$

$$y = G \cdot \begin{bmatrix} -5 \\ 5 \end{bmatrix} = (-5 + 5 \cdot t) \cdot e^{3t}$$

The point we are making is that we found a solution to the DE almost completely by matrix arithmetic. This point of view could be employed in very general terms to support much of what we did with constant coefficient DE's.

3. Problems

10.1. Let P_2 be the vector space of polynomials of degree at most 2. Compute $G^{-1} \cdot (2 \cdot t + 3 \cdot t^2)$ where G is the coordinate transformation for each of the bases listed. (Thus, you will compute a different $G^{-1} \cdot (2 \cdot t + 3 \cdot t^2)$ each time.)
a)   Use the basis $1, t, t^2$;
b)   Use the basis $1 + t, 1 - t, t + t^2$;
c)   Use the basis $1 - t, t + t^2, 1 + t$.

10.2. Let $f : U \to V$ and $g : U \to V$ be linear transformations (so we are asserting that U, V are vector spaces). Define the function $f + g$ in the usual way: $(f + g)(u) = f \cdot u + g \cdot u$ for all u in U. Show that $f + g$ is a linear transformation.

10.3. Let $f : U \to V$ be a linear transformation and let α be a number. Define $\alpha \cdot f$ as a function from U to V by $(\alpha \cdot f) \cdot u = \alpha \cdot (f \cdot u)$ for all u in U. Show that $\alpha \cdot f$ is a linear transformation. (Note: this problem and the previous one go a long way toward the following fact: the set of linear transformations from vector space U to vector space V forms a vector space!)

10.4. Let P_3 be the set of polynomials of degree at most 3, and P_2 those of degree at most 2. Let $D : P_3 \to P_2$ be differentiation, and let $J : P_2 \to P_3$ be defined by $J \cdot f(t) = \int_1 f(t) \cdot dt$. Show that $D \cdot J \cdot f(t) = f(t)$ for all $f(t)$ on P_2. Find $g(t)$ in P_2 such that $J \cdot D \cdot g(t) \neq g(t)$.

10.5. (Continuing the previous problem.) Use the natural bases for P_3 and P_2, and find the matrix $\hat{D}$ that represents D and the matrix $\hat{J}$ that represents J. Show that $\hat{D} \cdot \hat{J} = I_3$ and that $\hat{J} \cdot \hat{D} \neq I_4$.

10.6. Let V be the vector space of functions $(a + b \cdot t + c \cdot t^2) \cdot e^{3t}$, where a, b, c are arbitrary real numbers. Find a matrix that represents taking the derivative on V.

10.7. Consider the operator polynomial $Q = D^2 + D - 6$. Let V be the vector space of functions of the form $(a + b \cdot t) \cdot e^t$, where a, b are arbitrary real numbers. Show that if v is in V, then $Q \cdot v$ is in V. Find a representing matrix for Q and show that it has an inverse. Use the matrix inverse to solve the equation

$$y'' + y' - 6 \cdot y = t \cdot e^t \quad \text{which is} \quad Q \cdot y = t \cdot e^t$$

10.8. Let $p_0(x)$, $p_1(x)$, $p_2(x)$ be the Legendre polynomials found on page 131. Use these polynomials as a basis for the space P_2 of polynomials of degree at most 2. Let $D : P_2 \to P_2$ be the differential operator, as usual. Find a matrix A that represents D in this context. Show that $A^3 = \mathbb{O}_{3 \times 3}$. (This last equation says that the third derivative of a quadratic polynomial is 0.)

10.9. Let $r : \mathbb{R}^2 \to \mathbb{R}^2$ be reflection about the line $y = -2 \cdot x$. Find the 2×2 matrix A such that $A \cdot p = r \cdot p$ for all p in $\mathbb{R}^2$. Show that $A^2 = I_2$.

10.10. Let V_3 be the vector space of functions $(a + b \cdot t + c \cdot t^2) \cdot e^{-t}$, where a, b, c are arbitrary real numbers. Let V_2 be the vector space of functions $(a + b \cdot t) \cdot e^{-t}$. Show that the operator polynomial $D + 1$ maps V_3 to V_2. Using the natural bases of V_3 and V_2, find the matrix that represents $D + 1$.

10.11. Let $L : U \to V$ be a linear transformation. Define W to be the set of u in U such that $L \cdot u = \mathbb{O}_V$. The set W is called the *kernel* of L. Show that W is a subspace of V. (How do you tell that a subset is a subspace? A proposition on p.141 is relevant.)

CHAPTER 11

Eigenvalues

We are entering deeper waters. It is a profound insight that eigenvalues give the solution to many linear algebra problems. This is one of those topics that is better appreciated in hindsight – after you have seen the various uses. We will concentrate at first on the basic definitions and algorithms. It is at this point that we really need to allow complex number scalars, for non-real numbers come up naturally even in problems that involve real measurements. For instance, we have seen that non-real exponents arise naturally in the solutions of real differential equations. When complex entries are allowed in matrices and vectors, the definition and properties of matrix addition, multiplication, and scalar multiplication are unchanged. We still have the identity matrices and the zero matrices with their characteristic properties. The definition of *vector space* is the same, except that complex number scalars are allowed. The algorithms for calculating basis and dimension are the same.

For an $n \times n$ matrix A (with complex entries!) and a complex number λ and a non-zero $n \times 1$ matrix V, if we have the equation

$$(11.1) \qquad\qquad A \cdot V = \lambda \cdot V$$

then we say that λ is an *eigenvalue* of A and that V is an *eigenvector*. To link λ and V, we say that V *belongs to* λ.

The terms just introduced involve the German prefix "eigen," which means "own" or "characteristic." Thus, we can call λ a *characteristic value* of A and we can say that V is a *characteristic vector*. The standard terminology sticks *mostly* to the German.

1. Computing Eigenvalues and Their Vectors

A little work on equation (11.1) will disclose the whole story of eigenvalues, but first we give a couple of very simpleminded examples. If A is an arbitrary $n \times n$ matrix, then we have

$$A \cdot \mathbb{O}_{n \times 1} = \mathbb{O}_{n \times 1} = \alpha \cdot \mathbb{O}_{n \times 1} \quad \text{for every number} \quad \alpha$$

If we allowed $\mathbb{O}_{n \times 1}$ to be an eigenvector, then every number would be an eigenvalue of every matrix. While this might seem nicely inclusive, it gets in the way of most applications, and so we do not allow V in (11.1) to be the zero matrix.

Eigen-*vectors* cannot be 0, but eigen-*values* (the λ) can be 0, so that the equation

$$\begin{pmatrix} 2 & -2 \\ -2 & 2 \end{pmatrix} \cdot \begin{pmatrix} 1 \\ 1 \end{pmatrix} = \begin{pmatrix} 0 \\ 0 \end{pmatrix} = 0 \cdot \begin{pmatrix} 1 \\ 1 \end{pmatrix}$$

identifies $\begin{pmatrix} 1 \\ 1 \end{pmatrix}$ as an eigenvector of $\begin{pmatrix} 2 & -2 \\ -2 & 2 \end{pmatrix}$ with eigenvalue 0.

We have $I_n V = 1 \cdot V$ for every $n \times 1$ matrix V. Thus, 1 is an eigenvalue for I_n and every non-zero V is an eigenvector. Similarly familiar,

$$\mathbb{O}_{n \times n} \cdot V = \mathbb{O}_{n \times 1} = 0 \cdot V$$

for every $n \times 1$ matrix V, and so 0 is an eigenvalue for $\mathbb{O}_{n \times n}$. For real numbers a, b compute

$$\begin{pmatrix} a & 0 \\ 0 & b \end{pmatrix} \cdot \begin{pmatrix} u \\ v \end{pmatrix} = \begin{pmatrix} au \\ bv \end{pmatrix}$$

Thus, with $u = 1$ and $v = 0$, we see that a is an eigenvalue, and with $u = 0$, $v = 1$, we notice that b is an eigenvalue.

Starting with equation (11.1), we want to ask two questions: Given a matrix A, which numbers λ are eigenvalues for it? Given an eigenvalue, which vectors are eigenvectors?

The equation (11.1) can be written like this:

$$A \cdot V = \lambda \cdot I_n \cdot V$$

Subtracting, we find

$$\mathbb{O}_{n \times 1} = \lambda \cdot I_n \cdot V - A \cdot V \quad \text{or} \quad \mathbb{O}_{n \times 1} = (\lambda \cdot I_n - A) \cdot V$$

Consider the system of linear equations

$$(11.2) \qquad\qquad (\lambda \cdot I_n - A) \cdot X = \mathbb{O}_{n \times 1}$$

The set of solutions to this equation is called the λ-*eigenspace of A*. Because of the zero right side, an eigenspace is a vector space. The *non-zero* elements of the eigenspace are the λ-eigenvectors. Such non-zero vectors exist if and only if the system (11.2) has more than one solution. This happens if and only if the matrix $(\lambda \cdot I_n - A)$ has rank less than n, if and only if that matrix is **not invertible**, if and only if the matrix has *determinant zero*. We have arrived at the following.

PROPOSITION 11.1. *Let A be an $n \times n$ matrix and $\lambda \in \mathbb{C}$. Then λ is an eigenvalue of A if and only if $\det(\lambda \cdot I_n - A) = 0$. If λ is an eigenvalue of A, then v is an eigenvector belonging to λ if and only if v is a non-zero solution to the equation $(\lambda \cdot I_n - A) \cdot X = \mathbb{O}_{n \times 1}$.*

Let's see what Proposition 11.1 looks like in the case of a 2×2 matrix. Let

$$A = \begin{pmatrix} a & b \\ c & d \end{pmatrix}$$

Then

$$\lambda \cdot I_2 - A = \begin{pmatrix} \lambda - a & -b \\ -c & \lambda - d \end{pmatrix}$$

Taking the determinant and simplifying, we obtain

$$\det(\lambda \cdot I_2 - A) = \lambda^2 - (a + d) \cdot \lambda + ad - bc$$

We see that λ is an eigenvalue if and only if it is a root of the polynomial

$$X^2 - (a + d)X + ad - bc$$

This work is suggestive: λ is an eigenvalue of the 2×2 matrix given in the first problem if and only if it is the root of a polynomial formed from that matrix. We generalize this to an arbitrary $n \times n$ matrix A. It turns out that $\det(\lambda \cdot I_n - A)$ is a polynomial in λ of degree n; it is called the *characteristic polynomial of A*.

We have reduced the problem of finding eigenvalues to the problem of finding the roots of a polynomial. You probably know that solving polynomial equations in general is a tricky business, even if all you are interested in is numerical approximation. In introducing the subject, we will try to keep the polynomials fairly simple to facilitate computation, but you should realize that eigenvalues can be very difficult to compute, even difficult to approximate!

Consider the matrix

$$M = \begin{pmatrix} 0 & 0 & -1 \\ 1 & 1 & 0 \\ -2 & 2 & 2 \end{pmatrix}$$

Compute

$$\begin{pmatrix} \lambda & 0 & 0 \\ 0 & \lambda & 0 \\ 0 & 0 & \lambda \end{pmatrix} - \begin{pmatrix} 0 & 0 & -1 \\ 1 & 1 & 0 \\ -2 & 2 & 2 \end{pmatrix} = \begin{pmatrix} \lambda & 0 & 1 \\ -1 & \lambda - 1 & 0 \\ 2 & -2 & \lambda - 2 \end{pmatrix}$$

and the determinant (the characteristic polynomial) comes out like this:

$$\lambda^3 - 3 \cdot \lambda^2 + 4$$

We can factor this polynomial (mostly trial and error) to get

$$\lambda^3 - 3\lambda^2 + 4 = (\lambda + 1)(\lambda - 2)^2$$

and this shows that the eigenvalues of M are -1 and 2.

Each eigenvalue has its own set of eigenvectors, and so we find the eigenvectors separately for each eigenvalue, using Proposition 11.1 each time. For eigenvalue -1, the eigenvalues are the non-zero solutions to the equation $(-I_3 - M) \cdot X = \mathbb{O}$ (we have substituted -1 in for λ). We have

$$-I_3 - M = \begin{pmatrix} -1 & 0 & 1 \\ -1 & -2 & 0 \\ 2 & -2 & -3 \end{pmatrix}$$

and we solve the system whose augmented matrix is this:

$$\begin{pmatrix} -1 & 0 & 1 & 0 \\ -1 & -2 & 0 & 0 \\ 2 & -2 & -3 & 0 \end{pmatrix}$$

Elimination gives solutions

(11.3) $x_1 = x_3, \quad x_2 = -(1/2)x_3, \quad x_3 \quad$ free

The form (11.3) shows that the (-1)-eigenspace has dimension 1 with basis

$$\begin{bmatrix} 1 \\ -1/2 \\ 1 \end{bmatrix}$$

Any non-zero value of x_3 in (11.3) will give a non-zero solution which is then an eigenvector for the eigenvalue -1. For example, if $x_3 = 2$, we get solution $x_1 = 2, x_2 = -1, x_3 = 2$. To test this, you should check that equation (11.1), the defining equation of eigenvalues and vectors, holds in this case.

$$\begin{pmatrix} 0 & 0 & -1 \\ 1 & 1 & 0 \\ -2 & 2 & 2 \end{pmatrix} \cdot \begin{pmatrix} 2 \\ -1 \\ 2 \end{pmatrix} = (-1) \cdot \begin{pmatrix} 2 \\ -1 \\ 2 \end{pmatrix}$$

For the eigenvalue 2, we solve the system $(2 \cdot I_3 - M) \cdot X = \mathbb{O}_{3 \times 1}$. We have

$$2 \cdot I_3 - M = \begin{pmatrix} 2 & 0 & 1 \\ -1 & 1 & 0 \\ 2 & -2 & 0 \end{pmatrix}$$

The solutions are:

$$x_1 = -(1/2)x_3, \quad x_2 = -(1/2)x_3, \quad x_3 \quad \text{free}$$

The 2-eigenspace has dimension 1 and basis

$$\begin{bmatrix} -1/2 \\ -1/2 \\ 1 \end{bmatrix}$$

For instance, $x_1 = -1, x_2 = -1, x_3 = 2$ gives a particular eigenvector. Formulate and check the validity of equation (11.1) in this context.

Because eigenvalues are the roots of polynomials, they can easily be non-real, complex numbers. Recall the rotation matrix

$$R(\pi/2) = \begin{bmatrix} 0 & -1 \\ 1 & 0 \end{bmatrix}$$

The characteristic polynomial is $\lambda^2 + 1$ with roots $\lambda = \pm i$, where $i^2 = -1$. We can find the eigenvectors as we did before, using complex numbers:

$$i \cdot I_2 - R(\pi/2) = \begin{bmatrix} i & 1 \\ -1 & i \end{bmatrix} \quad \text{with eigenvectors} \quad x_2 \cdot \begin{bmatrix} i \\ 1 \end{bmatrix}$$

The $-i$ eigenvectors are found in the same way. In Chapter 7 we saw that complex roots occur naturally; so with eigenvalues, and in Chapter 12, the complex eigenvalues will be used in the solution of differential equations.

We close this chapter with a theoretical result of importance. Let A be an $n \times n$ matrix, and let $f(t)$ be a polynomial. We might write

$$f(t) = a_0 + a_1 \cdot t + a_2 \cdot t^2 + \cdots + a_k \cdot t^k$$

where $a_0, a_1, \ldots, a_k$ are constants. We want to define what it means to plug A into the polynomial; it means this:

$$f(A) = a_0 \cdot I_n + a_1 \cdot A + a_2 \cdot A^2 + \cdots + a_k \cdot A^k$$

In other words, t^j becomes A^j exactly as in the case of plugging in a number. The only wrinkle is the a_0 term, which we sometimes think of as $a_0 \cdot 1$; when we plug in A, we use I_n for "1." We see that $f(A)$ is a sum of $n \times n$ matrices, and so it is $n \times n$, itself.

Here is an approach to eigenvalues that uses matrix polynomials.

PROPOSITION 11.2. *Let A be an $n \times n$ matrix. Then there is a non-zero polynomial $f(t)$ such that $f(A) = \mathbb{O}$. If $f(t)$ is the such a polynomial of degree as small as possible, then λ is an eigenvalue of A if and only if $f(\lambda) = 0$.*

PROOF. The matrices $I_n, A, A^2, A^3, \ldots$ are vectors in the finite dimensional vector space of $n \times n$ matrices.[1] Therefore, they cannot be linearly independent, and there are scalars $a_0, a_1, \ldots, a_m$ for some m, not all 0, with

$$a_0 \cdot I_n + a_1 \cdot A + \cdots + a_m \cdot A^m = \mathbb{O}_{n \times n}$$

Define $f(t) = a_0 + a_1 \cdot t + \cdots + a_m \cdot t^m$ and we have a non-zero polynomial with $f(A) = \mathbb{O}$.

Let $f(t)$ be a non-zero polynomial with $f(A) = \mathbb{O}$ and having degree as small as possible. We claim that the roots of $f(t)$ are exactly the eigenvalues of A. Let λ be an eigenvalue of A, and find an eigenvector v belonging to λ. A homework exercise shows that $A \cdot v = \lambda \cdot v$ leads to

$$f(A) \cdot v = f(\lambda) \cdot v$$

Since $f(A) = \mathbb{O}$, we see that $f(\lambda) \cdot v = \mathbb{O}$. The eigenvector is not 0, and therefore $f(\lambda) = 0$. In other words, all A's eigenvalues are roots of $f(t)$.

---

[1]The space of $n \times n$ matrices has dimension n^2.

Conversely, let $f(\lambda) = 0$ for some complex number λ. Then we can factor $f(t) = (t - \lambda) \cdot g(t)$ where $g(t)$ is a polynomial of degree less than $f(t)$. In terms of matrices, we have

$$\mathbb{O} = f(A) = (A - \lambda \cdot I_n) \cdot g(A)$$

Since $f(t)$ has minimal degree such that $f(A) = \mathbb{O}$, we see that $g(A) \neq \mathbb{O}$. There is therefore an $n \times 1$ matrix v such that $g(A) \cdot v \neq \mathbb{O}$. Define $w = g(A) \cdot v$, and we claim that w is an eigenvector for A with eigenvalue λ. Compute

$$\mathbb{O} = (A - \lambda \cdot I_n) \cdot g(A) \cdot v = (A - \lambda \cdot I_n) \cdot w = A \cdot w - \lambda \cdot w$$

This shows that $\lambda \cdot w = A \cdot w$, as needed. $\square$

A polynomial of minimal degree for the matrix A as in Proposition 11.2 is called the *minimal polynomial of A*. Its roots are the eigenvalues of A. To compute the minimal polynomial, you can consider I_n, A, then I_n, A, A^2, then I_n, A, A^2, A^3, and so on, until you come to a linear dependence. Here are some examples if you want to experiment.

Examples of minimal polynomials.

matrix:	I_n	$\mathbb{O}_{n \times n}$	$\begin{bmatrix} -1 & 3 \\ 4 & -2 \end{bmatrix}$	$\begin{bmatrix} 1 & 1 \\ 0 & 1 \end{bmatrix}$
mimimal polynomial:	$t - 1$	t	$t^2 + 3 \cdot t - 10$	$t^2 - 2 \cdot t + 1$

It is also a fact that the minimal polynomial is a factor of the characteristic polynomial. It follows that the degree of the minimal polynomial of an $n \times n$ matrix is less than or equal to n.

2. Problems

11.1. Use the polynomial formula on p.178 to find the eigenvalues of the following matrices. Then find the eigenvectors that belong to those eigenvalues. (As noted, complex eigenvalues occur naturally.)

a) $\begin{pmatrix} 1 & 2 \\ 2 & -2 \end{pmatrix}$ **b)** $\begin{pmatrix} 0 & 1 \\ 2 & 0 \end{pmatrix}$ **c)** $\begin{pmatrix} 1 & 1 \\ 0 & 1 \end{pmatrix}$ **d)** $\begin{pmatrix} 2 & 4 \\ -1 & 2 \end{pmatrix}$

11.2. Find the characteristic polynomial, eigenvalues and a basis for each of the eigenspaces for these matrices.

a) $\begin{pmatrix} 2 & -1 & 3 \\ 0 & 2 & 1 \\ 0 & 0 & 4 \end{pmatrix}$ **b)** $\begin{pmatrix} 0 & 0 & 1 \\ 0 & 1 & 0 \\ 1 & 0 & 0 \end{pmatrix}$ **c)** $\begin{pmatrix} -1 & 6 & 7 \\ 0 & 0 & 2 \\ -1 & 6 & 3 \end{pmatrix}$

11.3. Find the characteristic polynomial, eigenvalues and and a basis for each of the eigenspaces for these matrices. (Note: The characteristic polynomial for (b) is $\lambda^4 - 3 \cdot \lambda^2 - 2 \cdot \lambda$.)

a) $\begin{pmatrix} 3 & 1 & 2 \\ 0 & 3 & 0 \\ 0 & 1 & -1 \end{pmatrix}$ **b)** $\begin{pmatrix} -1 & 0 & 0 & 0 \\ 1 & 5 & 3 & -3 \\ 1 & 3 & 1 & -1 \\ 1 & 9 & 5 & -5 \end{pmatrix}$

11.4. Show that the eigenvalues of the rotation matrix $R(\theta)$ are $\exp(\pm i \cdot \theta)$.

11.5. Show that the entries on the diagonal of a triangular matrix are the eigenvalues.

11.6. Let A be $n \times n$, $\lambda \in \mathbb{C}$, let v be $n \times 1$, and suppose that $A \cdot v = \lambda \cdot v$. Show that $A^j \cdot v = \lambda^j \cdot v$ for each positive integer j.

11.7. Let $f(t)$ be a polynomial, let A be $n \times n$, let λ be an eigenvalue for A, and let v be an eigenvector belonging to λ. Show that $f(A) \cdot v = f(\lambda) \cdot v$.

11.8. Let A be an $n \times n$ matrix. Then 0 is an eigenvalue of A if and only if A is *not invertible*.

11.9. Let $n \geq 2$ and let A be $n \times n$ with every entry 1. Show that $0, n$ are eigenvalues of A. (Hint: find eigenvectors explicitly.)

11.10. Use that $\det(A^T) = \det(A)$ to show that A and A^T have the same characteristic polynomial. (Thus, they have the same eigenvalues.)

11.11. Suppose that A is $n \times n$ and that each of its columns adds up to the same number λ. Let J be $n \times 1$ having all its entries equal to 1, and compute $A^T \cdot J$ to identify an eigenvalue of A^T. Conclude that A has the same eigenvalue.

11.12. Recall Leontief's model of an economy on p.66, and the idea of a Markov process. Let A be an $n \times n$ matrix, each of whose columns adds up to 1. Then there is a non-zero $n \times 1$ matrix P such that $A \cdot P = P$.

11.13. Recall the matrix
$$M = \begin{pmatrix} 0 & 1 \\ 1 & 1 \end{pmatrix}$$
associated in Chapter 4 with the Fibonacci sequence. Find an eigenvalue λ for M. Show that $\lambda^{n+2} = \lambda^{n+1} + \lambda^n$. (Hint: use the characteristic polynomial that gives λ as a root.)

11.14. Let A be $n \times n$ and let $f(t)$ be a polynomial such that $f(A) = \mathbb{O}$. Show that every eigenvalue of A is a root of $f(t)$. (Hint: a previous problem involved $f(A) \cdot v$ where v is an eigenvector for eigenvalue λ.)

11.15. Let $L : V \to V$ be a linear transformation (note that the vector space V is mapped to itself). For a real number λ, we say that λ is an *eigenvalue* of L if there is a non-zero vector v such that $L \cdot v = \lambda \cdot v$. Let A be a matrix that represents L. Show that A's real number eigenvalues are eigenvalues of L.

11.16. Let V be the vector space spanned by e^{2t}, $t \cdot e^{2t}$, and let $D : V \to V$ be differentiation, considered as a linear transformation. Find the eigenvalues and eigenvectors for D.

CHAPTER 12

Linear Systems of Differential Equations

1. First and Second Order Problems

We come to the culmination of our work on DE's. This chapter involves a generalization of our work on constant coefficient equations in Chapter 7 and on eigenvalues in Chapter 11. We want to work with an $n \times 1$ matrix X of functions of t. Thus,

$$X = \begin{pmatrix} X_1(t) \\ X_2(t) \\ \vdots \\ X_n(t) \end{pmatrix}$$

Of course, each X_j is the matrix entry $X[j, 1]$, but we will prefer the subscript notation in this chapter. We define the derivative of X by taking the derivative of each of its entries:

$$X' = \begin{pmatrix} X_1'(t) \\ X_2'(t) \\ \vdots \\ X_n'(t) \end{pmatrix}$$

We are interested in the following type of problem: given an $n \times n$ matrix A, solve

$$(12.1) \qquad\qquad X' = A \cdot X$$

This has a long name: it is a *homogeneous linear system of constant coefficient DE's*. If we multiply out the matrices on the right, we see that this is actually n differential equations in the functions $X_j(t)$. The DE associated with row j

of A is this:

$$X'_j = A[j,1] \cdot X_1 + A[j,2] \cdot X_2 + \cdots + X_n \cdot A[j,n]$$

where the entries of A are constants. The matrix A is the *coefficient matrix*.

An initial value problem for (12.1) consists of $X(0)$, which is the $n \times 1$ matrix consisting of the $X_j(0)$. Since we are using subscript notation for the X_j, we will always write the 0 value of t in function notation: $X_j(0)$.

A *solution* to (12.1) is an $n \times 1$ matrix X of specific functions that satisfies the equation. Notice that the set of solutions is a vector space! Here is an abstract description of that space, thinking that (12.1) identifies multiplication by A as a linear transformation. We are looking for a vector space of functions on which multiplication by A represents differentiation.

There is a close relation between the problem (12.1) and the constant co-efficient problems considered in Chapter 7. Similar to the case of the earlier chapter, we have scaling and superposition in the present case. Those principles repeat what we said before about the set of solutions being a vector space.

Principle of Scaling   If X is a solution to (12.1) and if a is a real number, then $a \cdot X$ is a solution to the equation.

Principle of Superposition.   If U, V are solutions to (12.1), then so is $U + V$.

Example.   Recall the most famous DE.

$$x' = -y$$
$$y' = x$$

We can write this in matrix form.

$$\frac{d}{dt}\begin{pmatrix} x \\ y \end{pmatrix} = \begin{pmatrix} x' \\ y' \end{pmatrix} = \begin{pmatrix} -y \\ x \end{pmatrix} = \begin{pmatrix} 0 & -1 \\ 1 & 0 \end{pmatrix} \cdot \begin{pmatrix} x \\ y \end{pmatrix}$$

so the equation is

$$X' = \begin{pmatrix} 0 & -1 \\ 1 & 0 \end{pmatrix} \cdot X$$

(It is interesting that the coefficient matrix is the rotation matrix for the angle $\pi/2$; more on this later.) ■

It turns out that every constant coefficient DE can be turned into a linear system. Thus, this present chapter is actually a generalization of Chapter 7. We won't need to work with this in general, but here's an example for what it's worth.

Problem. Write the following DE as a linear system of DE's.

$$y^{(3)} + y'' - 4 \cdot y' - 4 \cdot y = 0 \quad \text{and} \quad y_0 = -1, \ y_0' = 3, \ y_0'' = 0$$

Solution. The trick is to use new functions for the derivatives of y, up to where the DE kicks in. Define $X_1 = y$, $X_2 = y'$, $X_3 = y''$, and compute

$$\begin{pmatrix} X_1 \\ X_2 \\ X_3 \end{pmatrix}' = \begin{pmatrix} X_1' \\ X_2' \\ X_3' \end{pmatrix} = \begin{pmatrix} y' \\ y'' \\ y^{(3)} \end{pmatrix}$$

The first two derivatives of y can be written in terms of the X_j; the third derivative uses the DE.

$$\begin{pmatrix} y' \\ y'' \\ y^{(3)} \end{pmatrix} = \begin{pmatrix} X_2 \\ X_3 \\ -y'' + 4 \cdot y' + 4 \cdot y \end{pmatrix} = \begin{pmatrix} X_2 \\ X_3 \\ -X_3 + 4 \cdot X_2 + 4 \cdot X_1 \end{pmatrix}$$

and so we end up the this system of linear DE's.

$$\begin{pmatrix} X_1 \\ X_2 \\ X_3 \end{pmatrix}' = \begin{pmatrix} 0 & 1 & 0 \\ 0 & 0 & 1 \\ 4 & 4 & -1 \end{pmatrix} \cdot \begin{pmatrix} X_1 \\ X_2 \\ X_3 \end{pmatrix}$$

There is an equation involving second derivatives that looks very much like (12.1):

(12.2) $$X'' = A \cdot X$$

where, as before, X is an $n \times 1$ matrix of functions of t and A is an $n \times n$ matrix of numbers. Because of the occurrence of X'', this is called a *second order problem*.

We can relate this equation to (12.1) using a trick similar to the one employed in the problem above. Write

$$Y = \begin{pmatrix} X \\ X' \end{pmatrix}$$

and then

$$Y' = \begin{pmatrix} X' \\ X'' \end{pmatrix} = \begin{pmatrix} X' \\ A \cdot X \end{pmatrix} \begin{pmatrix} \mathbb{O} \cdot X + I_n \cdot X' \\ A \cdot X + \mathbb{O} \cdot X' \end{pmatrix} = \begin{pmatrix} \mathbb{O} & I_n \\ A & \mathbb{O} \end{pmatrix} \cdot \begin{pmatrix} X \\ X' \end{pmatrix}$$

so that

(12.3) $$Y' = \begin{pmatrix} \mathbb{O} & I_n \\ A & \mathbb{O} \end{pmatrix} \cdot Y$$

The equation (12.2) can be solved by solving (12.3); we will see an example in a later section.

In the next three sections, we will discuss techniques for solving systems of linear DE's.

2. The Eigenvalue/Eigenvector Method

The E/E Method.   *Let A be an $n \times n$ matrix, let λ be an eigenvalue, and let V be an eigenvector. Then $X = \exp(\lambda \cdot t) \cdot V$ is a solution to $X' = AX$.*

To see that this works, notice that since V is $n \times 1$, the expression $X = \exp(\lambda \cdot t) \cdot V$ is also $n \times 1$, and its entries are functions of t, since $\exp(\lambda \cdot t)$ multiplies each of the constant entries in V. It is easy to take the derivative of X.

$$X' = \lambda \cdot \exp(\lambda \cdot t) \cdot V$$

The quantities λ and $\exp(\lambda \cdot t)$ are scalar, and so they commute with matrices. Thus, we can write

$$X' = \lambda \cdot V \cdot \exp(\lambda \cdot t) = A \cdot V \cdot \exp(\lambda \cdot t) = A \cdot X$$

In other words, X is a solution to (12.1): $X' = A \cdot X$.

Problem. Solve the most famous IVP

$$X' = \begin{pmatrix} 0 & -1 \\ 1 & 0 \end{pmatrix} \cdot X \quad \text{and} \quad X_0 = \begin{pmatrix} 1 \\ 0 \end{pmatrix}$$

Solution. To get solutions, we need the eigenvalues and eigenvectors for the coefficient matrix. We have already calculated these. The eigenvalues are $\pm i$, and here are eigenvectors for them.

$$i: \begin{pmatrix} 1 \\ -i \end{pmatrix} \qquad -i: \begin{pmatrix} 1 \\ i \end{pmatrix}$$

Each eigenvalue/eigenvector gives a solution as above:

$$\exp(i \cdot t) \cdot \begin{pmatrix} 1 \\ -i \end{pmatrix} \quad \text{and} \quad \exp(-i \cdot t) \cdot \begin{pmatrix} 1 \\ i \end{pmatrix}$$

By scaling, constant multiples of these solutions are also solutions. By superposition, sums of solutions are solutions. Thus, we arrive at a general solution

$$X = A_1 \cdot \exp(i \cdot t) \cdot \begin{pmatrix} 1 \\ -i \end{pmatrix} + A_2 \cdot \exp(-i \cdot t) \cdot \begin{pmatrix} 1 \\ i \end{pmatrix}$$

The initial conditions:

$$\begin{pmatrix} 1 \\ 0 \end{pmatrix} = X_0 = A_1 \cdot \begin{pmatrix} 1 \\ -i \end{pmatrix} + A_2 \cdot \begin{pmatrix} 1 \\ i \end{pmatrix} = \begin{pmatrix} 1 & 1 \\ -i & i \end{pmatrix} \cdot \begin{pmatrix} A_1 \\ A_2 \end{pmatrix}$$

A system of linear equations! (By now, we are never surprised by linear equations.) We can solve: $A_1 = 1/2$ and $A_2 = 1/2$, and we get a specific solution.

Putting that solution into one matrix (with a little fiddling) gives us the solution we are expecting.

$$X = \frac{1}{2} \cdot \exp(i \cdot t) \cdot \begin{pmatrix} 1 \\ -i \end{pmatrix} + \frac{1}{2} \cdot \exp(-i \cdot t) \cdot \begin{pmatrix} 1 \\ i \end{pmatrix} = \begin{pmatrix} (e^{it} + e^{-it})/2 \\ i \cdot (-e^{it} + e^{-it})/2 \end{pmatrix}$$

$$= \begin{pmatrix} (e^{it} + e^{-it})/2 \\ i^2 \cdot (-e^{it} + e^{-it})/(2i) \end{pmatrix} = \begin{pmatrix} (e^{it} + e^{-it})/2 \\ (e^{it} - e^{-it})/(2i) \end{pmatrix} = \begin{pmatrix} \cos(t) \\ \sin(t) \end{pmatrix}$$

Here is a typical problem, involving the present chapter along with Chapter 11. The eigenvalues have been made workable for hand calculation.

Problem. Solve the IVP

$$X' = \begin{pmatrix} 21 & 42 & -63 \\ 25 & 44 & -63 \\ 26 & 48 & -70 \end{pmatrix} \cdot X \quad \text{and} \quad X(0) = \begin{pmatrix} 2 \\ 6 \\ 5 \end{pmatrix}$$

Solution. We need the eigenvalues and eigenvectors of the coefficient matrix. The characteristic polynomial can be computed with the aid of a calculator.

$$\det \begin{pmatrix} \lambda - 21 & -42 & 63 \\ -25 & \lambda - 44 & 63 \\ -26 & -48 & \lambda + 70 \end{pmatrix}$$

$$= (\lambda - 21) \cdot [\lambda^2 + 26\lambda - 56] + 42 \cdot [-25\lambda - 112] + 63 \cdot [26\lambda + 56]$$

$$= \lambda^3 + 5 \cdot \lambda^2 - 14 \cdot \lambda = \lambda \cdot (\lambda + 7) \cdot (\lambda - 2)$$

Now the eigenvectors that go with each eigenvalue. The eigenvalue is listed at the top of its column.

$$\begin{bmatrix} 0 & -7 & 2 \\ \hline -1 & 3 & 0 \\ 2 & 1 & 3 \\ 1 & 2 & 2 \end{bmatrix}$$

We get a general solution

$$X = a_1 \cdot \begin{pmatrix} -1 \\ 2 \\ 1 \end{pmatrix} + a_2 \cdot e^{-7t} \cdot \begin{pmatrix} 3 \\ 1 \\ 2 \end{pmatrix} + a_3 \cdot e^{2t} \cdot \begin{pmatrix} 0 \\ 3 \\ 2 \end{pmatrix}$$

The intial values:

$$\begin{pmatrix} 2 \\ 6 \\ 5 \end{pmatrix} = a_1 \cdot \begin{pmatrix} -1 \\ 2 \\ 1 \end{pmatrix} + a_2 \cdot \begin{pmatrix} 3 \\ 1 \\ 2 \end{pmatrix} + a_3 \cdot \begin{pmatrix} 0 \\ 3 \\ 2 \end{pmatrix}$$

and this has (convenient!) solution $a_1 = 1 = a_2 = a_3$. Thus, the solution to our IVP is this:

$$X = \begin{pmatrix} -1 \\ 2 \\ 1 \end{pmatrix} + e^{-7t} \cdot \begin{pmatrix} 3 \\ 1 \\ 2 \end{pmatrix} + e^{2t} \cdot \begin{pmatrix} 0 \\ 3 \\ 2 \end{pmatrix}$$

∎

The eigenvalue/eigenvector method is best used when the eigenvalues and vectors for the coefficient matrix are known or easy to get. It is also an important fact that $X = \exp(\lambda \cdot t) \cdot V$ is a solution to $X' = AX$ for every eigenvalue λ of A and every eigenvector V for λ. However, the eigenvector/eigenvalue method does not solve all IVP's for some matrices. The trouble is related to the repeated root case back in Chapter 7. Here is an example.

Example.   Consider the IVP

$$X' = \begin{pmatrix} 0 & 1 \\ 0 & 0 \end{pmatrix} \cdot X \quad \text{and} \quad X(0) = \begin{pmatrix} 0 \\ 1 \end{pmatrix}$$

The characteristic polynomial is λ^2, and so 0 is the only eigenvalue. The eigenvectors are multiples of $\begin{pmatrix} 1 \\ 0 \end{pmatrix}$, and so we get solutions to the DE of the form

$$X = A_1 \cdot \begin{pmatrix} 1 \\ 0 \end{pmatrix}$$

We see that we *cannot* solve the IVP, since we cannot get $X_2(0)$ to equal 1 in this solution.

In the next section we will show how to solve all linear systems of constant coefficient IVP's.

3. The Exponential Method

This method gives both theoretical and numerical information. Given the $n \times n$ matrix A and the real number t, we define

$$\exp(A \cdot t) = \sum_{k=0}^{\infty} \frac{t^k}{k!} \cdot A^k$$

This formula borrows from the Taylor series for the exponential function. Each term is the product of the scalar $t^k/k!$ and the $n \times n$ matrix A^k, and so each term is an $n \times n$ matrix. We will be somewhat informal about convergence,[1] merely asserting that the series produces an $n \times n$ matrix of functions. It is also permissible to differentiate the series term by term, so we have

$$\frac{d}{dt} \exp(A \cdot t) = \sum_{k=1}^{\infty} k \cdot \frac{t^{k-1}}{k!} \cdot A^k$$

$$= A \cdot \sum_{k=1}^{\infty} \frac{t^{k-1}}{(k-1)!} \cdot A^{k-1}$$

$$= A \cdot \exp(A \cdot t)$$

The A factor on the exponential can be pulled out on the right side as well, and we have this:

$$(12.4) \qquad \frac{d}{dt} \exp(A \cdot t) = A \cdot \exp(A \cdot t) = \exp(A \cdot t) \cdot A$$

In other words, we differentiate $\exp(A \cdot t)$ just as we differentiate the function $\exp(\alpha \cdot t)$ where α is a numerical constant.

Problem. Use (12.4) to show that if A is $n \times n$, then

$$\exp(A \cdot t) \cdot \exp(-A \cdot t) = I_n$$

---

[1]Technically, we need to estimate the radius of convergence of the sequence defined by each entry of the matrix partial sums. This is ugly but not really hard. We'll skip the details in this course.

Solution. We use (12.4) and the product rule.[2]

$$\frac{d}{dt}\Big[\exp(A \cdot t) \cdot \exp(-A \cdot t)\Big]$$

$$= \Big[\frac{d}{dt}\exp(A \cdot t)\Big] \cdot \exp(-A \cdot t) + \exp(A \cdot t) \cdot \Big[\frac{d}{dt}\exp(-A \cdot t)\Big]$$
$$= A \cdot \exp(A \cdot t) \cdot \exp(-A \cdot t) + \exp(A \cdot t) \cdot (-A) \cdot \exp(-A \cdot t)$$
$$= A \cdot \exp(A \cdot t) \cdot \exp(-A \cdot t) - A \cdot \exp(A \cdot t) \cdot \exp(-A \cdot t) = \mathbb{O}$$

The derivative of $\exp(A \cdot t) \cdot \exp(-A \cdot t)$ is the zero matrix. Thus, every entry of the product matrix is a constant. We can find the constant by plugging in $t = 0$, and we get $\exp(\mathbb{O}) \cdot \exp(\mathbb{O}) = I_n$, as needed. ■

We show how to use the exponential to solve the differential equation (12.1). Suppose that $X(0)$ is a given $n \times 1$ matrix, and define

(12.5) $$X(t) = \exp(A \cdot t) \cdot X(0)$$

Observe that

$$X'(t) = A \cdot \exp(A \cdot t) \cdot X(0) = A \cdot X$$

and we see that X is a solution to the system of linear DE's! This is directly analogous to solving $y' = k \cdot y$ by the ordinary exponential $y = y_0 \cdot e^{kt}$, except that in the present case, we have to put the $X(0)$ on the right so that we get a meaningful expression for X. Continuing that analogy, let's show that (12.5) gives the *unique* solution to the DE. Indeed, suppose that

$$Y'(t) = A \cdot Y(t)$$

---

[2]The product rule holds for products of matrices with function entries. You have to keep the matrices in the same order throughout.

Just as in the case of the real function $e^{\alpha \cdot t}$, we multiply Y by $\exp(-A \cdot t)$ and use the product rule:

$$\frac{d}{dt}\big[\exp(-A \cdot t) \cdot Y(t)\big] = \exp'(-A \cdot t) \cdot Y(t) + \exp(-A \cdot t) \cdot Y'(t)$$
$$= \exp(-A \cdot t) \cdot (-A) \cdot Y(t) + \exp(-A \cdot t) \cdot A \cdot Y(t)$$
$$= \mathbb{O}$$

The derivative of the matrix $\exp(-A \cdot t) \cdot Y(t)$ is the zero matrix. That means that every entry in $\exp(-A \cdot t) \cdot Y(t)$ is constant. Let C be that constant. Then

$$\exp(-A \cdot t) \cdot Y(t) = C$$

Using the identity proved above: $\exp(A \cdot t) \cdot \exp(-A \cdot t) = I_n$, we multiply both sides of the $Y(t)$ equation on the left by $\exp(A \cdot t)$, and we get

$$Y(t) = \exp(A \cdot t) \cdot C$$

Plugging in $t = 0$, we see that $Y(0) = C$, and we have uniqueness.

Here is a very simple example. We could not use the eigenvalue/eigenvector method to solve

$$X' = \begin{pmatrix} 0 & 1 \\ 0 & 0 \end{pmatrix} \cdot X \quad \text{and} \quad X(0) = \begin{pmatrix} 0 \\ 1 \end{pmatrix}$$

It is simple to use (12.5) to do so. Let A be the 2×2 coefficient matrix in this problem, and then

$$X = \sum_{k=0}^{\infty} \frac{t^k}{k!} \cdot A^k \cdot \begin{pmatrix} 0 \\ 1 \end{pmatrix}$$

Compute that $A^2 = \mathbb{O}_{2 \times 2}$, and so $A^k = \mathbb{O}$ for all $k \geq 2$. We get

$$X = [I_2 + t \cdot A] \cdot \begin{pmatrix} 0 \\ 1 \end{pmatrix} = \begin{pmatrix} 1 & t \\ 0 & 1 \end{pmatrix} \cdot \begin{pmatrix} 0 \\ 1 \end{pmatrix} = \begin{pmatrix} t \\ 1 \end{pmatrix}$$

and we have the unique solution to the IVP.

In general, the powers of a matrix A aren't $\mathbb{O}$, and so we would need to be clever to get an exact formula for $\exp(A \cdot t)$.

We mentioned that the exponential formula can be used numerically. Indeed, cutting off the infinite series at some finite point, we obtain

$$X(t) \approx \left(\sum_{k=0}^{m} \frac{t^k}{k!} \cdot A^k \right) \cdot X(0)$$

for each positive integer m. In general, the series converges very quickly, and we can get very good approximations without too much trouble. You might notice that the approximation on the right has polynomial entries.

4. Non-homogeneous Systems

Just as in Chapter 7, we can consider *non-homogeneous* problems:

$$X' = A \cdot X + B(t)$$

where A is the $n \times n$ coefficient matrix and $B(t)$ is an $n \times 1$ column of functions.

A principal from Chapter 7 survives: given a specific solution Y to the equation, and given a solution U to the homogeneous equation $U' = AU$, the matrix $Y + U$ gives a solution to the non-homogeneous equation. Here is a oft-occurring general example: If $B(t) = B$ is constant and A is invertible, then notice that $Y = -A^{-1} \cdot B$ is a constant (equilibrium) solution to the system. If we can solve $U' = A \cdot U$, then

$$U - A^{-1} \cdot B$$

will be a more general solution.

The exponential method lends itself to a generalization of the technique for First Order Linear DE's, introduced back in Chapter 2: we claim that

$$(12.6) \qquad X(t) = \exp(A \cdot t) \cdot \left[\int_0 \exp(-A \cdot t) \cdot B(t) \cdot dt + X(0) \right]$$

The proof is straightforward, but we will not indulge. Notice that the integrand is an $n \times 1$ matrix of functions; the integral is obtained by integrating each function separately and then forming the vector of integrals. This formula can be used in several ways, not the least of which is to obtain numerical approximations, as we described above.

In the case that A is invertible and $B(t) = B$ is constant, the formula (12.6) simplifies to this:

$$(12.7) \qquad X(t) = \exp(A \cdot t) \cdot \left[X(0) + A^{-1} \cdot B \right]$$

5. Applications

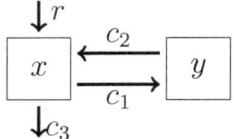

5.1. A biochemistry problem. Albumin occurs as an abundant blood plasma protein and also in extravascular fluids (lymph and tissue fluids). (See [15].) It is important in regulating blood volume and for other reasons. The picture represents the production and diffusion of albumin. When albumin is synthesized in the liver, it enters the vascular system (the square on the left) at a constant rate r. Albumin in the vascular system enters the extravascular system (the square on the right) at a relative rate of c_1 of the amount in the vascular system. Extravascular albumin enters the vascular system at a relative rate c_2 of the amount in the extravascular fluids. Albumin is excreted from the vascular fluids at a relative rate c_3. We will show that there is a stable equilibrium (no matter what the proportionality constants are).[3]

Let $x(t)$ be the amount in vascular fluids and $y(t)$ the amount in extravascular fluids. Then

$$x' = r - c_1 \cdot x + c_2 \cdot y - c_3 \cdot x$$
$$y' = c_1 \cdot x - c_2 \cdot y$$

---

[3]In a typical individual, $r \approx 9\text{g/day}$, $c_1 \approx 0.05$.

The related homogeneous equation:

$$\begin{aligned} x' &= -(c_1 + c_3) \cdot x + c_2 \cdot y \\ y' &= c_1 \cdot x - c_2 \cdot y \end{aligned} \quad \text{which is} \quad \begin{pmatrix} x \\ y \end{pmatrix}' = \begin{pmatrix} -(c_1 + c_3) & c_2 \\ c_1 & -c_2 \end{pmatrix} \cdot \begin{pmatrix} x \\ y \end{pmatrix}$$

The characteristic polynomial of the coefficient matrix is

$$\lambda^2 + (c_1 + c_2 + c_3) \cdot \lambda + c_2 \cdot c_3$$

As shown in Chapter 7, the fact that this has positive coefficients shows that its roots (the eigenvalues!) have negative real part and so their exponentials decay to 0 – thus, the solution U to the homogeneous equation is the typical transient.

If Y is an equilibrium solution to the original equation, then, as remarked above, $Y + U$ is the general solution to the original equation. Since $U \to \mathbb{O}$ as $t \to \infty$, the solution approaches the equilibrium, and that proves that the equilibrium is stable. ■

5.2. Springs on a line. The picture depicts two masses connected by three springs between fixed walls on the left and right. The circles are masses m_1 and m_2, respectively, and the line segments represent springs. The leftmost spring has spring constant k_0 and natural length h_0, the other constants give corresponding information for the other two springs. Suppose that the left wall is at $x = 0$, that mass m_1 is at $x_1(t)$ and that m_2 is at $x_2(t)$. Let's find an equilibrium for x_1, x_2. If we are not at equilibrium, we show that x_1 and x_2 oscillate. (We assume that the mass m_1 stays to the left of m_2.)

As in Chapter 7, we know that the force on a mass exerted by a spring is proportional to the displacement from the natural length, and the constant of proportionality is the *spring constant*. The force on m_1 comes from springs 0 and 1: from spring 0 the force is $-k_0 \cdot (x_1 - h_0)$ and from spring 1 the force is $-k_1 \cdot (h_1 - (x_2 - x_1))$. In other words

$$m_1 \cdot x_1'' = -k_0 \cdot (x_1 - h_0) - k_1 \cdot (h_1 - (x_2 - x_1))$$

so that

$$(12.8) \qquad x_1'' = -\frac{k_0 + k_1}{m_1} \cdot x_1 + \frac{k_1}{m_1} \cdot x_2 + \frac{k_0 \cdot h_0 - k_1 \cdot h_1}{m_1}$$

Similarly, the force on m_2 comes from springs 1 and 2. Let c be the position of the right wall.

$$m_2 \cdot x_2'' = -k_1 \cdot (x_2 - x_1 - h_1) - k_2 \cdot (x_2 - (c - h_2))$$

and this is

$$(12.9) \qquad x_2'' = \frac{k_1}{m_2} \cdot x_1 - \frac{k_1 + k_2}{m_2} \cdot x_2 + \frac{k_1 \cdot h_1 + k_2 \cdot (c - h_2)}{m_2}$$

The resulting system is an example of a non-homogeneous version of the second order system (12.2):

$$\begin{pmatrix} x_1 \\ x_2 \end{pmatrix}'' = \begin{pmatrix} -(k_0 + k_1)/m_1 & k_1/m_1 \\ k_1/m_2 & -(k_1 + k_2)/m_2 \end{pmatrix} \cdot \begin{pmatrix} x_1 \\ x_2 \end{pmatrix}$$
$$+ \begin{pmatrix} (k_0 \cdot h_0 - k_1 \cdot h_1)/m_1 \\ (k_1 \cdot h_1 + k_2 \cdot (c - h_2))/m_2 \end{pmatrix}$$

We get equilibrium in this system by setting $x_1'' = 0 = x_2''$ and solving the resulting equations for x_1, x_2. Given specific values for the coefficients, the equations have a solution, since the determinant of the 2×2 coefficient matrix is not 0.

As for the general system, just as above, a solution can be formed as $Y + U$ where Y is a given solution to the non-homogeneous equation and U is a solution to the related homogeneous equation. For a solution to the non-homogeneous equation, we can let Y be the equilibrium. To get U, we use the equation (12.3), obtained for a general second order problem:

$$U' = \begin{pmatrix} 0 & 0 & 1 & 0 \\ 0 & 0 & 0 & 1 \\ -(k_0 + k_1)/m_1 & k_1/m_1 & 0 & 0 \\ k_1/m_2 & -(k_1 + k_2)/m_2 & 0 & 0 \end{pmatrix} \cdot U$$

We will show that U oscillates by showing that the eigenvalues of the coefficient matrix have the form $b \cdot i$ where b is real. The characteristic polynomial of the coefficient matrix takes a little work; here it is.

$$\lambda^4 + \left(\frac{k_0 + k_1}{m_1} + \frac{k_1 + k_2}{m_2} \right) \cdot \lambda^2 + \frac{k_0 \cdot k_1 + k_0 \cdot k_2 + k_1 \cdot k_2}{m_1 \cdot m_2}$$

The quadratic formula and some messy algebra(!) show that λ^2 has two, negative real values, and so the four eigenvalues have the form $b \cdot i$ where b is real. Thus, U can be put into cosine/sine form, and it oscillates. ∎

6. Problems

12.1. Solve the following IVP, using the eigenvalue/eigenvector method.

$$\frac{d}{dt}\begin{pmatrix} x_1 \\ x_2 \end{pmatrix} = \begin{pmatrix} 2 & 1 \\ 8 & -5 \end{pmatrix} \cdot \begin{pmatrix} x_1 \\ x_2 \end{pmatrix} \quad \text{and} \quad \begin{pmatrix} x_1(0) \\ x_2(0) \end{pmatrix} = \begin{pmatrix} 3 \\ -6 \end{pmatrix}$$

12.2. Solve the following IVP, using the eigenvalue/eigenvector method.

$$\frac{d}{dt}\begin{pmatrix} y_1 \\ y_2 \end{pmatrix} = \begin{pmatrix} 1 & -1 \\ 5 & -1 \end{pmatrix} \cdot \begin{pmatrix} y_1 \\ y_2 \end{pmatrix} \quad \text{and} \quad \begin{pmatrix} y_1(0) \\ y_2(0) \end{pmatrix} = \begin{pmatrix} 6 \\ 0 \end{pmatrix}$$

12.3. Solve the following IVP, using the eigenvalue/eigenvector method.

$$\frac{d}{dt}\begin{pmatrix} X_1 \\ X_2 \\ X_3 \end{pmatrix} = \begin{pmatrix} 8 & -2 & 4 \\ -2 & 3 & 4 \\ 3 & 1/2 & 4 \end{pmatrix} \cdot \begin{pmatrix} X_1 \\ X_2 \\ X_3 \end{pmatrix} \quad \text{and} \quad X(0) = \begin{pmatrix} 10 \\ 10 \\ 10 \end{pmatrix}$$

12.4. Show that
$$\exp\left(\begin{bmatrix} 1 & 1 \\ 0 & 1 \end{bmatrix} \cdot t\right) = \begin{bmatrix} e^t & t \cdot e^t \\ 0 & e^t \end{bmatrix}$$

(Hint: the powers of the matrix inside the exponential come out nicely! Write the series out term by term to see the pattern.)

12.5. Use the previous problem and the exponential method to solve the following IVP. (Note that the eigenvalue-eigenvector method will not work on this problem.)
$$X' = \begin{pmatrix} 1 & 1 \\ 0 & 1 \end{pmatrix} \cdot X \quad \text{and} \quad X(0) = \begin{pmatrix} 0 \\ 1 \end{pmatrix}$$

12.6. In the albumin model discussed above, replace the constant rate r by an oscillating rate: $74 \cdot \sin(2t)$, where t is time in days. Assume that $c_1 = 1$, $c_2 = 2$, $c_3 = 3$. Find a specific solution of the form $x = A_1 \cdot \cos(2t) + B_1 \cdot \sin(2t)$ and $y = A_2 \cdot \cos(2t) + B_2 \cdot \sin(2t)$, where A_1, B_1, A_2, B_2 are constants. (The general solution will decay toward this specific solution.)

12.7. In the model of the two masses held by three springs, let $m_1 = 1$, $m_2 = 2$, $k_0 = 1$, $k_1 = 4$, $k_2 = 2$, $h_0 = 1$, $h_1 = 2$, $h_2 = 1$. Find the equilibrium. (The answer will involve c.) Show that the force of spring 0 on mass 1 is the same as the force of spring 1 on mass 2.

12.8. We have three chemicals A, B, C in a common medium. Substance A turns into substance B at a rate proportional to the amount of A; substance B turns into C at a rate proportional to the amount of B; substance C turns into A at a rate proportional to the amount of C. Find a system of DE's describing the amounts of each chemical. Show that 0 is an eigenvalue of the coefficient matrix, and show that its eigenvectors are equilibria.

12.9. The picture on the next page depicts an energy grid, as in the problem on p.70. The numbers A, B are constant temperatures, and the x_j change in time. The rate of x_j is equal to the sum of the differences $y - x_j$ where y is a node connected to x_j. (So y can be A or B or one of the other x_k.) Write down the system of DE's that arises. You might be interested in the eigenvalues of the coefficient matrix; we suggest using numerical software to estimate them. You will see that they are negative, and it follows that the solutions drift toward equilibrium.

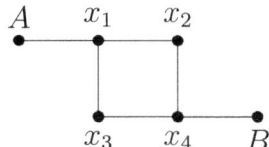

CHAPTER 13

Diagonalizable Matrices

The theoretical part of this subject gives important insights into its applications. However, in our course we have time only to give a quick introduction. Therefore, we will be as practical as possible. We will defer explaining where this idea comes up, focussing rather on the terminology and computation needed to categorize a given matrix.

Here's is the key definition: an $n \times n$ matrix A is *diagonalizable* if every $n \times 1$ vector is a linear combination of eigenvectors of A. We feel free to abbreviate the word *diagonalizable* to *d'ble*.

We begin with two examples. The matrix

$$A = \begin{pmatrix} 1 & 1 \\ 0 & 1 \end{pmatrix}$$

has characteristic polynomial $(\lambda - 1)^2$ and so 1 is its only eigenvalue. The eigenvectors for 1 are these:

$$\begin{pmatrix} a \\ 0 \end{pmatrix} = a \cdot \begin{pmatrix} 1 \\ 0 \end{pmatrix}$$

Thus, the span of the eigenvectors of A is the set of scalar multiples of a single matrix. In particular, if V is 2×1 and $V[2, 1] \neq 0$, then V cannot be in the span of the eigenvector of A. In other words, A is *not diagonalizable*.

On the other hand, notice that

$$I_n \cdot V = 1 \cdot V \quad \text{for every} \quad n \times 1 \quad \text{matrix} \quad V$$

Thus, every $n \times 1$ matrix is an eigenvector of I_n, and so this matrix is d'ble.

1. Determining that a Matrix is Diagonalizable

We present two methods, each with its own advantages and disadvantages. Recall the definition of the minimal polynomial $f(t)$ of the matrix A: the non-zero polynomial of smallest degree with $f(A) = \mathbb{O}$. We showed that the eigenvalues of A are exactly the roots of $f(t)$. The proof that the following test works takes us a little too far afield in algebra to warrant giving the argument in this course.

Diagonalizable Test 1. The matrix A is diagonalizable if and only if its minimal polynomial has no repeated roots. ■

Our second test takes a little more set-up; it involves the characteristic polynomial $f(\lambda) = \det(\lambda \cdot I_n - A)$ of A. Let α be an eigenvalue of A (so that $f(\alpha) = 0$). Then $f(\lambda)$ can be factored

$$f(\lambda) = (\lambda - \alpha)^m \cdot g(\lambda)$$

where $g(\alpha) \neq 0$. The number m is called the *algebraic multiplicity of* α. The eigenvectors for α are the non-zero elements of the null space of $\alpha \cdot I_n - A$. The dimension k of this null space is called the *geometric multiplicity of* α. It is always the case that $k \leq m$.

Diagonalizable Test 2. The matrix A is diagonalizable if the algebraic multiplicity is equal to the geometric multiplicity for every eigenvalue of A. ■

Problem. Let

(13.1)
$$A = \begin{pmatrix} 0 & 0 & 1 \\ 0 & 1 & 0 \\ 1 & 0 & 0 \end{pmatrix}$$

Solution. First up, Test 2: The characteristic polynomial is $(\lambda - 1)^2 \cdot (\lambda + 1)$. The eigenvalue 1 has algebraic multiplicity 2. Its geometric multiplicity is the dimension of the null space of

$$1 \cdot \begin{pmatrix} 1 & 0 & 0 \\ 0 & 1 & 0 \\ 0 & 0 & 1 \end{pmatrix} - \begin{pmatrix} 0 & 0 & 1 \\ 0 & 1 & 0 \\ 1 & 0 & 0 \end{pmatrix} = \begin{pmatrix} 1 & 0 & -1 \\ 0 & 0 & 0 \\ -1 & 0 & 1 \end{pmatrix}$$

The null space has dimension 2, equal to the algebraic multiplicity. We are not done, since we need to consider *all* the eigenvalues. The eigenvalue -1 remains; its algebraic multiplicity is 1 (from the characteristic polynomial). The geometric multiplicity must be at least 1, since there are eigenvectors and they are not zero. The geometric multiplicity is always less than or equal to the algebraic multiplicity, and we see that they must be equal. Thus, the matrix is d'ble.

We have the answer, but it's interesting to think about Test 1. Looking at $I_n, A, A^2, \ldots$, we find that

$$A^2 - I_3 = \mathbb{O}$$

and we see that the minimal polynomial is $t^2 - 1 = (t-1)(t+1)$. This polynomial has no repeated roots, and so A is d'ble. ∎

The argument we just gave for an eigenvector of algebraic multiplicity 1 always works; in this case the geometric multiplicity is 1 as well.

Diagonalizable One Way Test. If the characteristic polynomial of the matrix A has no repeated roots, then A is d'ble. ∎

In this case, the characteristic polynomial is equal to the minimal polynomial, so that Tests 1 and 2 coincide. The converse statement to the One Way Test is false: the characteristic polynomial can have repeated roots and A still be d'ble. We have seen an example in this chapter!

How to write a vector as a sum of eigenvectors. Recall that our definition of the $n \times n$ matrix A being d'ble is that all $n \times 1$ vectors are linear combinations of the eigenvectors of A. In the next section, we will see the uses of this. For now, we describe how to write a given vector as a sum of eigenvectors.

For each eigenvalue α of A, find a basis for the null space of $\alpha \cdot I_n - A$. Form a matrix P using all the basis vectors for all the eigenvalues. If A is d'ble, then P will be $n \times n$ and invertible. Given an $n \times 1$ matrix C, solve the equation $P \cdot X = C$ for its unique solution $X = E$. (We can use Elimination, or we might compute $E = P^{-1} \cdot C$.) If P_j is the j-th column of P for each j, then we have

$$C = \sum_{j=1}^{n} E[j] \cdot P_j$$

Each $E[j] \cdot P_j$ is either the zero vector, or it is an eigenvector of A.

Example. We return to the matrix A in the example (13.1). Here is a basis for the eigenvectors for each eigenvalue.

$$-1 : \begin{pmatrix} -1 \\ 0 \\ 1 \end{pmatrix} \qquad 1 : \begin{pmatrix} 1 & 0 \\ 0 & 1 \\ 1 & 0 \end{pmatrix}$$

Here is the matrix with these as columns:

$$P = \begin{pmatrix} -1 & 1 & 0 \\ 0 & 0 & 1 \\ 1 & 1 & 0 \end{pmatrix}$$

Since A is d'ble, every 3×1 matrix is a linear combination of eigenvectors of A. Let's try it out with the matrix that appears as the right side of the following system of equations.

$$\begin{pmatrix} -1 & 1 & 0 & | & 2 \\ 0 & 0 & 1 & | & 3 \\ 1 & 1 & 0 & | & -2 \end{pmatrix} \quad \text{has solution} \quad \begin{pmatrix} -2 \\ 0 \\ 3 \end{pmatrix}$$

and so we can write the vector as a sum of eigenvectors of A.

$$\begin{pmatrix} 2 \\ 3 \\ -2 \end{pmatrix} = -2 \cdot \begin{pmatrix} -1 \\ 0 \\ 1 \end{pmatrix} + 3 \cdot \begin{pmatrix} 0 \\ 1 \\ 0 \end{pmatrix} = \begin{pmatrix} 2 \\ 0 \\ -2 \end{pmatrix} + \begin{pmatrix} 0 \\ 3 \\ 0 \end{pmatrix}$$

2. Applications

Systems of DE's.   The eigenvalue/eigenvector method solves all IVP's for $X' = A \cdot X$ when A is d'ble.   Indeed, the initial vector $X(0)$ is a sum of eigenvectors, say

$$X(0) = \sum_{j=1}^{m} V_j$$

where V_j is an eigenvector for A, say with eigenvalue λ_j.   Then

$$X = \sum_{j=1}^{m} \exp(\lambda_j \cdot t) \cdot V_j$$

solves the IVP.

Problem.   Solve the IVP

$$\frac{dx}{dt} = -x + 3 \cdot y, \quad \frac{dy}{dt} = 4 \cdot x - 2 \cdot y, \quad x_0 = -13, \quad y_0 = 29$$

Solution.   Let

$$A = \begin{bmatrix} -1 & 3 \\ 4 & -2 \end{bmatrix} \quad \text{and} \quad X = \begin{bmatrix} x \\ y \end{bmatrix}$$

and our IVP is $X' = A \cdot X$ with $X(0) = \begin{bmatrix} -13 & 29 \end{bmatrix}^T$.

The characteristic polynomial for A is $t^2 + 3 \cdot t - 10 = (t+5) \cdot (t-2)$.   The One Way Test says that A is d'ble, and so we know that the eigenvalue/eigenvector method will win through.   Here is the matrix P whose first column is a basis for the -5 eigenspace and whose second column is a basis for the 2 eigenspace.

$$P = \begin{bmatrix} -3 & 1 \\ 4 & 1 \end{bmatrix}$$

We solve

$$P \cdot E = \begin{bmatrix} -13 \\ 29 \end{bmatrix} \quad \text{and get} \quad E = \begin{bmatrix} 6 \\ 5 \end{bmatrix}$$

This shows that

$$\begin{bmatrix} -13 \\ 29 \end{bmatrix} = 6 \cdot \begin{bmatrix} -3 \\ 4 \end{bmatrix} + 5 \cdot \begin{bmatrix} 1 \\ 1 \end{bmatrix}$$

and here is our solution.

$$\begin{bmatrix} x \\ y \end{bmatrix} = X = 6 \cdot e^{-5t} \cdot \begin{bmatrix} -3 \\ 4 \end{bmatrix} + 5 \cdot e^{2t} \cdot \begin{bmatrix} 1 \\ 1 \end{bmatrix}$$

∎

Turning a matrix into a diagonal matrix.   When A is $n \times n$ and d'ble, its eigenvectors span $\mathbb{R}^n$, and so there is a basis for $\mathbb{R}^n$ consisting of eigenvectors for A. Let P be $n \times n$ having as its columns such a basis. Then P is invertible, and if column j of P has eigenvalue λ_j, then we claim that the following formula holds.

$$P^{-1} \cdot A \cdot P = \begin{pmatrix} \lambda_1 & 0 & \cdots & 0 & 0 \\ 0 & \lambda_2 & \cdots & 0 & 0 \\ \vdots & \vdots & \ddots & \vdots & \vdots \\ 0 & 0 & \cdots & \lambda_{n-1} & 0 \\ 0 & 0 & \cdots & 0 & \lambda_n \end{pmatrix}$$

This can be verified by multiplying both sides on the left by P, and then comparing the j-th column on each side, for $1 \leq j \leq n$. The matrix on the right is said to be a *diagonal matrix*.

The diagonal matrix on the right allows us to do arithmetic with A rather easily. For example, for each positive integer k, we have

$$A^k = P \cdot \begin{pmatrix} \lambda_1^k & 0 & \cdots & 0 & 0 \\ 0 & \lambda_2^k & \cdots & 0 & 0 \\ \vdots & \vdots & \ddots & \vdots & \vdots \\ 0 & 0 & \cdots & \lambda_{n-1}^k & 0 \\ 0 & 0 & \cdots & 0 & \lambda_n^k \end{pmatrix} \cdot P^{-1}$$

(Two of the problems at the end of the chapter deal with this formula.)

Markov processes.   An example was given in Chapter 4. Suppose we are on one of two states at a given time in sequence. If we are in state 1 at one point in the sequence, suppose that the chance is 60% that we will be in state 1 at the next time. If we are in state 2, suppose that the chance is 90% that we will be in state 2 at the next time. The long-term behavior of the states is governed by the limit

$$\lim_{n \to \infty} \begin{pmatrix} 0.6 & 0.1 \\ 0.4 & 0.9 \end{pmatrix}^n$$

We can compute this limit if we change the matrix into a diagonal matrix.

The characteristic polynomial is $(\lambda - 1)(\lambda - 0.5)$ and so the matrix is d'ble. A basis for the eigenvectors for each eigenvalue: (1 first, 0.5 second).

$$\begin{pmatrix} 0.1 & -1 \\ 0.4 & 1 \end{pmatrix}$$

and so

$$\begin{pmatrix} 0.6 & 0.1 \\ 0.4 & 0.9 \end{pmatrix}^n = \begin{pmatrix} 0.1 & -1 \\ 0.4 & 1 \end{pmatrix} \cdot \begin{pmatrix} 1^n & 0 \\ 0 & 0.5^n \end{pmatrix} \cdot \begin{pmatrix} 0.1 & -1 \\ 0.4 & 1 \end{pmatrix}^{-1}$$

As $n \to \infty$, we have $1^n \to 1$ and $0.5^n \to 0$. Thus, the limit is

$$\begin{pmatrix} 0.1 & -1 \\ 0.4 & 1 \end{pmatrix} \cdot \begin{pmatrix} 1 & 0 \\ 0 & 0 \end{pmatrix} \cdot \begin{pmatrix} 0.1 & -1 \\ 0.4 & 1 \end{pmatrix}^{-1} = \begin{pmatrix} 0.1 & 0 \\ 0.4 & 0 \end{pmatrix} \cdot \begin{pmatrix} 2 & 2 \\ -0.8 & 0.2 \end{pmatrix}$$
$$= \begin{pmatrix} 0.2 & 0.2 \\ 0.8 & 0.8 \end{pmatrix}$$

This says that, in the long run, state 1 occurs 20% of the time, and state 2 occurs 80% of the time.

3. Problems

13.1. Go back to all the matrices for which we have computed eigenvalues and eigenvectors; determine which matrices are d'ble and which are not. (Note: this should be a matter of looking up the dimension of each eigenspace and comparing to the algebraic multiplicity.)

13.2. Let D be a diagonal matrix. Show, for each positive integer k, that D^k is diagonal, with $D^k[j, j] = (D[j, j])^k$ for each j.

13.3. Let P be an invertible $n \times n$ matrix and let E be an $n \times n$ matrix. Show, for each positive integer k, that

$$\left(P^{-1} \cdot E \cdot P\right)^k = P^{-1} \cdot E^k \cdot P$$

13.4. (Continuation of the previous problem.) Show that

$$P^{-1} \cdot \exp(E) \cdot P = \exp(P^{-1} \cdot E \cdot P)$$

13.5. (Continuation of the previous.) Let

$$E = \begin{pmatrix} -5 & 2 \\ 10 & 3 \end{pmatrix}$$

Use eigenvectors to find a 2×2 matrix P such that $P^{-1} \cdot E \cdot P$ is diagonal. Now use the previous problem to compute $\exp(E)$.

13.6. Let A be a d'ble $n \times n$ matrix such that $|\lambda| < 1$ for every eigenvalue λ of A. Show that $A^k \to \mathbb{O}_{n \times n}$ as $k \to \infty$. (Hint: Use the previous problem. Note: the limit holds even if A is not d'ble, but it's harder to prove.)

13.7. Recall that an $n \times n$ matrix A with real entries is *symmetric* if $A^T = A$. It turns out that every symmetric matrix is d'ble. Show that this is true for an arbitrary 2×2 matrix. (Hint: if the eigenvalues are distinct, the matrix is d'ble. How could there be a repeated eigenvalue?)

13.8. Let

$$F = \frac{1}{3} \cdot \begin{pmatrix} 1 & -4 & 4 \\ -2 & -1 & 4 \\ -2 & -4 & 7 \end{pmatrix}$$

Compute the limit of F^n as $n \to \infty$. (Hint: if λ is an eigenvalue of M, then $\lambda/3$ is an eigenvalue of $M/3$, and $M/3$ has the same eigenvectors as $M/3$.)

CHAPTER 14

Linear Partial Differential Equations

1. Some Famous PDE's

Recall that if we have a function $U(x, y)$ of two independent variables x, y, then we define the *partial derivative* $\partial U / \partial x$ by taking the derivative in x while holding y fixed. In limits

$$\frac{\partial U}{\partial x}(x, y) = \lim_{h \to 0} \frac{U(x + h, y) - U(x, y)}{h}$$

Similarly,

$$\frac{\partial U}{\partial y}(x, y) = \lim_{h \to 0} \frac{U(x, y + h) - U(x, y)}{h}$$

We compute partial derivatives using all the usual differentiation rules, only remembering to hold the non-derivative variables constant.

It is customary to use a subscript notation for the partial derivative. We write

$$U_x = \frac{\partial U}{\partial x} \quad \text{and} \quad U_y = \frac{\partial U}{\partial y}$$

Similarly with higher order derivatives; here is the second derivative in x:

$$U_{xx} = \frac{\partial^2 U}{(\partial x)^2} = \frac{\partial}{\partial x} \cdot \frac{\partial U}{\partial x}$$

You may recall *mixed* partial derivatives, such as

$$U_{xy} = \frac{\partial}{\partial y} \cdot \frac{\partial U}{\partial x}$$

indicating that we first take the derivative in x, and then in y. Mixed partials really do occur! It turns out that if U_{xy} is continuous, then it is the same as the mixed partial in the other order: $U_{xy} = U_{yx}$. We repeat that these partial derivatives have to be continuous to be equal.

A *partial differential equation*, abbreviated *PDE*, is an equation in a function of several variables, involving the partial derivatives of that function. Differential equations in functions of only one variable are called *ordinary differential equations* to distinguish them from the partial kind. And the distinction is quite stark. Whereas ordinary differential equations are studied in *families*, such as the *linear differential equations* that we have studied extensively, each individual partial differential equation displays its own unique nuances in its solutions, and, generally speaking, the available solution techniques are specialized. We will introduce some of the most commonly occurring PDE's, give a very general solution technique (there are no *universal* techniques), and solve some representative problems. We will see that the concept of an *initial value problem* becomes more complicated. Our work will involve a brief digression into *sine series*, a species of *Fourier series* – a deep and rich subject.

1.1. The One-Dimensional Diffusion Equation. This equation is also called the *heat equation*.[1] There are versions of it for n-dimensional objects, for all $n \geq 1$; here we give the one-dimensional version. We have a function $U(x, t)$ where x is position along an object (let's say a steel bar) lying along the closed interval $[0, L]$, and $t \geq 0$ is time. The function $U(x, t)$ describes the energy (e.g. heat) at position x on the bar and at time t. The word *diffusion* implies that the energy at hotter areas of the bar diffuses to colder areas. At the molecular level, energy transfer is proportional to the difference in energy level, as stipulated by *Fourier's Law*. Here is the PDE:

$$(14.1) \qquad \frac{\partial U}{\partial t} = c^2 \cdot \frac{\partial^2 U}{(\partial x)^2}$$

---

[1]For a mathematical derivation, see [**2**, p.620-21].

The number c is a constant; we use c^2 to emphasize that the constant coefficient is positive.

1.2. The Two-Dimensional Diffusion Equation. The reader will see the obvious analogy with the one-dimensional case. We have the energy $U(x, y, t)$ at a point (x, y) on a planar object, and at time t. For instance, we might imagine that the object is a rectangle: $0 \leq x \leq L$ and $0 \leq y \leq M$. As before, we have $t \geq 0$.

$$(14.2) \qquad \frac{\partial U}{\partial t} = c^2 \cdot \left[\frac{\partial^2 U}{(\partial x)^2} + \frac{\partial^2 U}{(\partial y)^2} \right]$$

for a constant c.

1.3. The One-Dimensional Wave Equation. Waves occur in sound, light, water, and in many other situations of periodic vibrations. In the one-dimensional case, we have an object along the closed interval $[0, L]$, and $U(x, t)$ gives the *amplitude* of the vibration at point x and time t. This equation was considered very early on in the development of calculus by Euler and Daniel Bernoulli, especially as it concerns the vibration of strings on musical instruments.[2] The PDE:

$$(14.3) \qquad \frac{\partial^2 U}{(\partial t)^2} = c^2 \cdot \frac{\partial^2 U}{(\partial x)^2}$$

and c is a constant. There are versions of this equation in higher physical dimensions; the case of a two-dimensional drum is of particular interest to the musician.

1.4. Laplace's Equation. Laplace's equation involves functions $U(x, y)$ in two physical dimensions: the functions are called *harmonic functions*, and they occur in the study of electricity, gravity, fluid potential, heat conduction,

---

[2]For a derivation, see [**2**, p.631-2].

and in a number of other areas. For example, if $U(x, y, t)$ satisfies the two-dimensional heat equation, and if U is constant in time, then we can forget t and we have a harmonic function $U(x, y)$. Here is Laplace's PDE:

$$(14.4) \qquad \frac{\partial^2 U}{(\partial x)^2} + \frac{\partial^2 U}{(\partial y)^2} = 0$$

It is worth noting that this simple equation hides a far-reaching generality. There is a sense in which all the standard functions of mathematics give solutions to this equation. We will develop some inkling of this – the main point is that PDE's are not simple.

Laplace's equation has an important version in polar coordinates. If $U(x, y)$ satisfies Laplace's equation (if U is harmonic), we can define

$$W(r, \theta) = U(r \cdot \cos(\theta), r \cdot \sin(\theta)) \quad \text{for} \quad r \geq 0$$

Then, in subscript form,

$$(14.5) \qquad r^2 \cdot W_{rr} + r \cdot W_r + W_{\theta\theta} = 0$$

1.5. Schrödinger's Equation. This famous PDE describes how the quantum states of a quantum system change in time.[3] For a quantum particle somewhere on an interval $[0, L]$, we have a complex number valued function $\Psi(x, t)$, where x is position and t is time. The interpretation of Ψ is somewhat up in the air – its absolute value is usually taken to be a *probability density function* for the particle; we will not go into details, except to exhibit the equation:

$$\frac{ih}{2\pi} \cdot \frac{\partial \Psi}{\partial t} = -\frac{h^2}{8\pi^2 m} \cdot \frac{\partial^2 \Psi}{(\partial x)^2} + V \cdot \Psi$$

where $i^2 = -1$ and h is Planck's constant and m is the mass of the particle and V is the particle's potential energy.

---

[3]Obviously we are falling into the difficult quantum mechanics. An introduction to Schrödinger's equation is in [**16**, Chapter 42].

1.6. Superposition, Scaling, Separation. The PDE's we have introduced are all *linear*; that means that the set of solutions to each one forms a vector space. In more basic terms, the equations satisfy *superposition* and *scaling*. The demonstration of this for the individual equations is straightforward.

Problem. Let $U(x,t)$ and $V(x,t)$ be solutions to the diffusion equation (14.1). Then $U + V$ is also a solution. If α is a constant, then $\alpha \cdot U$ is a solution.

Solution. Compute

$$\frac{\partial}{\partial t}\Big(U + V\Big) = \frac{\partial U}{\partial t} + \frac{\partial V}{\partial t} = c^2 \cdot \frac{\partial^2 U}{(\partial x)^2} + c^2 \cdot \frac{\partial^2 V}{(\partial x)^2}$$

$$= c^2 \cdot \left(\frac{\partial^2 U}{(\partial x)^2} + \frac{\partial^2 V}{(\partial x)^2}\right) = c^2 \cdot \frac{\partial^2}{(\partial x)^2}\Big(U + V\Big)$$

and

$$\frac{\partial}{\partial t}\Big(\alpha \cdot U\Big) = \alpha \cdot \frac{\partial U}{\partial t} = \alpha \cdot c^2 \cdot \frac{\partial^2 U}{(\partial x)^2} = c^2 \cdot \frac{\partial^2}{(\partial x)^2}\Big(\alpha \cdot U\Big)$$

∎

All the PDE's we have introduced exhibit superposition and scaling of solutions, and, just as for ordinary differential equations, this means that if we find individual solutions, then we can form more general solutions as linear combinations of those individual solutions.

How can we find individual solutions? We will introduce one general technique, and although it will work in almost all the PDE's we are considering, it is not a panacea. The technique is called *separation of variables*. Let's see it in action.

Problem.   Find a solution to the diffusion equation (14.1) of the form

$$U(x, t) = X(x) \cdot T(t)$$

where, as indicated, X is a function of x only, and T is a function of t only. (Having x and t *separated* is the idea of separation of variables.)

Solution.   Hold t constant and compute $U_x = X'(x) \cdot T(t)$, and so $U_{xx} = X''(x) \cdot T(t)$. Similarly, $U_t = X \cdot T'$. The diffusion equation is then

$$X \cdot T' = c^2 \cdot X'' \cdot T$$

Let's divide by $X \cdot T$; it will be convenient to divide by c^2, too.[4]

$$\frac{1}{c^2} \cdot \frac{X \cdot T'}{X \cdot T} = \frac{X'' \cdot T}{X \cdot T} \quad \text{so that} \quad \frac{1}{c^2} \cdot \frac{T'}{T} = \frac{X''}{X}$$

The function T'/T is constant in x, and the function X''/X is constant in t; to make the two fractions equal, let's make them constant. In other words, let's suppose there is a number λ such that

$$\frac{1}{c^2} \cdot \frac{T'}{T} = \lambda \quad \text{and} \quad \frac{X''}{X} = \lambda$$

We get two ordinary differential equations, one in t and one in x:

$$T' = \lambda \cdot c^2 \cdot T \quad \text{and} \quad X'' = \lambda \cdot X$$

It is not hard to solve these problems using the methods of Chapter 7. We will do that with more specific information in the next section. For now, let's just write T_λ and X_λ for solutions to the two equations. Then

$$U_\lambda = X_\lambda \cdot T_\lambda$$

will be a particular solution to the diffusion equation. ∎

---

[4]A word about logic: we are trying to find a solution, and so $XT' = c^2 X''T$ is the *conclusion* to which we are working, not the *hypothesis* we are starting from. Thus, *dividing* by XT to get the next equation is logically multiplying that next equation to get the equation we divided. Separation of variables involves backwards reasoning.

Superposition and scaling apply to the solutions just obtained: if we have a finite set of numbers λ, and a constant A_λ for each one, then

$$U = \sum_\lambda A_\lambda \cdot X_\lambda \cdot T_\lambda$$

will be a solution to the diffusion equation.

2. Boundary Value Problems

The diffusion equation imagines variables x, t, where $0 \le x \le L$, for a constant L, and $t \ge 0$. We can plot the values (x, t) in the x, t-plane; we have a rectangular strip extending upwards from the interval $[0, L]$ on the x-axis. A typical problem for the diffusion equation specifies the values of the solution function U on the *boundary* of this strip: we need $U(x, 0)$ for $0 \le x \le L$, and $U(0, t)$ for $t \ge 0$, and $U(L, t)$ for $t \ge 0$. Specifying these values produces a *boundary value problem*. Such a problem corresponds to an initial value problem for an ordinary differential equation.

Let's see how separation of variables can be used to solve a particular boundary value problem associated with the diffusion equation.

Problem. (The heat decay problem.) Let L and c be positive constants. Let $f(x)$ be defined for $0 \le x \le L$, and assume also that $f(0) = 0$ and $f(L) = 0$. Find a function $U(x, t)$ such that

$$\frac{\partial U}{\partial t} = c^2 \cdot \frac{\partial^2 U}{(\partial x)^2}$$
$$U(x, 0) = f(x) \quad \text{for} \quad 0 \le x \le L$$
$$U(0, t) = 0 = U(L, t) \quad \text{for} \quad t \ge 0$$

∎

In the *heat decay problem* the energy is held constant at the two ends of the physical bar on which U measures temperature. We are given an initial heat distribution $f(x)$, and the heat equation takes over from there as time moves on. Physical considerations lead us to expect that U will gradually decay to 0, as the heat diffuses out the ends of the bar; that's why the problem is called the heat decay problem. Even though we know, more or less, what is going to happen, the solution is quite interesting because it elicits a *function representation problem* involving periodic oscillations.

Solution.   Separation of variables gave us solutions of the form

$$U_\lambda = X(x) \cdot T(t) \quad \text{where} \quad X'' = \lambda \cdot X \quad \text{and} \quad T' = c^2 \cdot \lambda \cdot T$$

To get the conditions $U(0,t) = 0 = U(L,t)$, we impose

$$X(0) = 0 = X(L)$$

You have dealt with this! It implies that $\lambda = -n^2 \cdot \pi^2/L^2$ for some integer n. Because we want X to be non-zero, we let n be a positive integer. Let $d = \pi/L$, and

$$X = b \cdot \sin(n \cdot d \cdot x)$$

for a constant b. Then, the equation in T has solution $T = \exp(-c^2 \cdot d^2 \cdot n^2 \cdot t)$. We end up with the following solution; since it depends on the integer n, we write $b = b_n$.

$$U_n = b_n \cdot \sin(n \cdot d \cdot x) \cdot \exp(-c^2 \cdot d^2 \cdot n^2 \cdot t)$$

We use superposition to put these solutions together to form one grand solution.

$$U(x,t) = \sum_{n=1}^{\infty} b_n \cdot \sin(n \cdot d \cdot x) \cdot \exp(-c^2 \cdot d^2 \cdot n^2 \cdot t)$$

We have taken the liberty of including *all* positive integers n, so that our formula is an infinite series. We will defer the question whether the series makes sense while we think about $U(x,0)$, which is supposed to be the function

$f(x)$. We want to have

(14.6)
$$f(x) = \sum_{n=1}^{\infty} b_n \cdot \sin(n \cdot d \cdot x)$$

(Remember that $d = \pi/L$.) The series on the right is a *sine series*, a form of *Fourier series*. The series writes $f(x)$ in terms of functions that are periodic ($2L$ is a period). It might surprise us to learn that $f(x)$ can be an arbitrary differentiable function, and there will be a formula (14.6) valid on the open interval $0 < x < L$. We will take this issue up briefly in the next section. ∎

Begging the question about the sine series, we solve another boundary value problem.

Problem. (*Laplace's Equation on a Rectangle.*) Let L, M be positive constants, and let $f(x)$ be a differentiable function on the closed interval $[0, L]$. We will be looking for a harmonic function $U(x, y)$ defined on the rectangle with $0 \le x \le L$ and $0 \le y \le M$. We want U to agree with $f(x)$ along the x-axis. Here is the problem:[5]

$$U_{xx} + U_{yy} = 0$$
$$U(x, 0) = f(x) \quad \text{for} \quad 0 < x < L$$
$$U(x, M) = 0 \quad \text{for} \quad 0 \le x \le L$$
$$U(0, y) = U(L, y) = 0 \quad \text{for} \quad 0 \le y \le M$$

Solution. Separation: $U = X \cdot Y$ and we have
$$\frac{X''}{X} + \frac{Y''}{Y} = 0$$

---

[5]This is an example of a *Dirichlet Problem* in which we seek for a harmonic function in a region, such that the function values are given on the boundary of the region.

Let $X(0) = 0 = X(L)$, and, as before, let $d = \pi/L$, and we get $X = b \cdot \sin(n \cdot d \cdot x)$. But this time

$$Y'' = n^2 \cdot d^2 \cdot Y$$

so that

$$Y = \alpha_n \cdot \exp(n \cdot d \cdot y) + \beta_n \cdot \exp(-n \cdot d \cdot y)$$

We want $U(x, 0) = f(x)$, and so we will let $Y(0) = 1$; we want $U(x, M) = 0$, and so we'll take $Y(M) = 0$. Equations:

$$\alpha_n + \beta_n = 1 \quad \text{and} \quad \alpha_n \cdot \exp(ndM) + \beta_n \cdot \exp(-ndM) = 0$$

This yields

$$\alpha_n = \frac{\exp(-ndM)}{\exp(-ndM) - \exp(ndM)} \quad \text{and} \quad \beta_n = \frac{-\exp(ndM)}{\exp(-ndM) - \exp(ndM)}$$

and our solution is

$$U(x, y) = \sum_{n=1}^{\infty} b_n \cdot \sin(n \cdot d \cdot x) \cdot [\alpha_n \cdot \exp(n \cdot d \cdot y) + \beta_n \cdot \exp(-n \cdot d \cdot y)]$$

Our specification that $\alpha_n + \beta_n = 1$ shows that

$$U(x, 0) = \sum_{n=1}^{\infty} b_n \cdot \sin(n \cdot d \cdot x)$$

We want this to be $f(x)$. We are in the same position as with the heat decay problem – we seek a sine series representation for $f(x)$. ∎

We begin to wonder whether all PDE's come down to separation of variables and Fourier series. Let's see a different approach. d'Alembert discovered a solution to the plucked string problem associated with the wave equation. The advantage of this solution is to proceed directly from the boundary values; the disadvantage is to hide the oscillations.

Problem. (*d'Alembert's solution to the plucked string*[6]) Let L, c be positive numbers, and let $f(x)$ be a twice differentiable function on $[0, L]$. Find a function $F(x)$, defined for all real numbers x, such that if

$$U(x, t) = F(x - c \cdot t) - F(-x - c \cdot t) \quad \text{for} \quad 0 \le x \le L \quad \text{and} \quad t \ge 0$$

then

$$U_{tt} = c^2 \cdot U_{xx}$$
$$U(x, 0) = f(x) \quad \text{for} \quad 0 < x < L$$
$$U(0, t) = 0 = U(L, t) \quad \text{for} \quad t \ge 0$$
$$U_t(x, 0) = 0 \quad \text{for} \quad 0 < x < L$$

Solution. Compute that

$$\frac{\partial^2}{(\partial t)^2} F(x - c \cdot t) = c^2 \cdot F''(x - c \cdot t) \quad \text{and} \quad \frac{\partial^2}{(\partial x)^2} F(x - c \cdot t) = F''(x - c \cdot t)$$

Thus, $F(x - c \cdot t)$ satisfies the wave equation. So does, $F(-x - c \cdot t)$, and, by superposition and scaling, their difference does, as well. Thus, $U_{tt} = c^2 \cdot U_{xx}$, no matter what $F(x)$ is, as long as $F''(x)$ exists.

We have $U(0, t) = F(-c \cdot t) - F(-c \cdot t) = 0$, again no matter what $F(x)$ is.

The condition $U(L, t) = 0$ is $F(L - c \cdot t) = F(-L - c \cdot t)$. Because $c \cdot t$ is an arbitrary number, this is precisely that $F(x)$ has period $2L$. Therefore, to define $F(x)$, we need only consider x on the interval $[-L, L]$.

The condition $U(x, 0) = f(x)$ is $F(x) - F(-x) = f(x)$, and we see that $F(-x) = F(x) - f(x)$. The function $f(x)$ is defined for $0 < x < L$, and then $F(-x)$ gives the value of F on the open interval $(-L, 0)$.

Next,

$$U_t(x, t) = -c \cdot F'(x - c \cdot t) + c \cdot F(-x - c \cdot t)$$

---

[6]See [**7**, p.41-2]

and so the equation $U_t(x, 0) = 0$ is

$$-c \cdot F'(x) + c \cdot F'(-x) = 0 \quad \text{so that} \quad F'(x) = F'(-x)$$

when $0 < x < L$. Integrating this equation, we see that $F(x) = -F(-x) + K$ for some constant K. Taking $x = t = 0$, we see from $U(0, 0) = 0$ that $F(0) = 0$, and so $K = 0$. Thus, $F(x) = -F(-x)$. Putting this together with $F(-x) = F(x) - f(x)$ yields

$$-F(x) = F(x) - f(x) \quad \text{so that} \quad \frac{1}{2} \cdot f(x) = F(x) \quad \text{for} \quad 0 < x < L$$

Then we define $F(x) = -F(-x) = -f(x)/2$ for $-L < x < 0$. Finally, we extend $F(x)$ to be $2L$ periodic.[7] ■

3. Fourier Series in Brief

We have seen how sine series arise in the solution of certain PDE's. This had been observed by many people by the middle of the 18th century, among them d'Alembert, Euler, the Bernoulli's, and Fourier.[8] In general, both sine and cosine are involved in formulas that look like this:

$$(14.7) \qquad f(x) = \frac{a_0}{2} + \sum_{k=1}^{\infty} (a_k \cdot \cos(k \cdot d \cdot x) + b_k \cdot \sin(k \cdot d \cdot x))$$

where the a_k and b_k are real numbers,[9] and $d = \pi/L$ where the domain of x is $[-L, L]$. The terms in the sum are thought of as *constituent frequencies* of f, and the form of (14.7) gives immediate information about any vibrations or oscillations associated naturally with f. Notice that the series has period $2\pi/d = 2L$.

Because the sine and cosine functions are periodic, the equation (14.7) looks like a fairly special formula. Around 1822, Fourier argued that *all* functions

---

[7]We have begged the question about the existence of the second derivative at the endpoints $0, L$. Think about this.

[8]For both an introduction and a deeper treatment of Fourier series, we recommend [**7**].

[9]There is a technical reason for dividing a_0 by 2.

defined on $[-\pi, \pi]$ have this form; in other words, the $f(x)$ on the left side of equation (14.7) can be *anything*. Although Fourier's claim is false in general, it is true quite generally – we will see an important case. In his honor, series as in equation (14.7) are called *Fourier series*. We should mention that despite three hundred years' worth of work on this problem, the full story of equation (14.7) is still not known.

We will be content to state three theorems giving Fourier series representations. Our first uses sine series. Notice that the domain of the function $f(x)$ is $[0, L]$; that's half the interval $[-L, L]$ of periodicity.

PROPOSITION 14.1. *Let $f(x)$ be differentiable on $[0, L]$. Let $d = \pi/L$. For each positive integer n, define*

$$b_n = \frac{2}{L} \int_0^L f(x) \cdot \sin(n \cdot d \cdot x) \cdot dx$$

Then

$$g(x) = \sum_{n=1}^{\infty} b_n \cdot \sin(n \cdot d \cdot x)$$

converges for all $x \in \mathbb{R}$. If $0 < x < L$, then $g(x) = f(x)$.

Next, a cosine series theorem, again using the interval $[0, L]$. The uniformity of the formulas for a_n here results in $a_0/2$ appearing in the infinite series.

PROPOSITION 14.2. *Let $f(x)$ be differentiable on $[0, L]$. Let $d = \pi/L$. For each non-negative integer n, define*

$$a_n = \frac{2}{L} \int_0^L f(x) \cdot \cos(n \cdot d \cdot x) \cdot dx$$

Then

$$g(x) = \frac{a_0}{2} + \sum_{n=1}^{\infty} a_n \cdot \cos(n \cdot d \cdot x)$$

converges for all $x \in \mathbb{R}$. If $0 \leq x \leq L$, then $g(x) = f(x)$.

And we can combine sine and cosine. This time, the interval is $[-L, L]$.

PROPOSITION 14.3. *Let $f(x)$ be differentiable on $[-L, L]$. Let $d = \pi/L$. For each non-negative integer n, define*

$$a_n = \frac{1}{L} \cdot \int_{-L}^{L} f(x) \cdot \cos(n \cdot d \cdot x) \cdot dx \quad and \quad b_n = \frac{1}{L} \cdot \int_{-L}^{L} f(x) \cdot \sin(n \cdot d \cdot x) \cdot dx$$

Then

$$g(x) = \frac{a_0}{2} + \sum_{n=1}^{\infty} \left[a_n \cdot \cos(n \cdot d \cdot x) + b_n \cdot \sin(n \cdot d \cdot x) \right]$$

converges for all $x \in \mathbb{R}$. If $-L < x < L$, then $g(x) = f(x)$. We have $g(L) = (f(L) + f(-L))/2$.

PROOF. For $0 \le x \le L$, define

$$h(x) = \frac{1}{2} \cdot \left[f(x) + f(-x) \right] \quad and \quad k(x) = \frac{1}{2} \cdot \left[f(x) - f(-x) \right]$$

We apply Proposition 14.2 to $h(x)$ and Proposition 14.1 to $k(x)$, both of which are differentiable on $[0, L]$.   □

4. Problems

14.1.   (One-dimensional diffusion with insulated ends.) Given L, c and $f(x)$, solve

$$U_t = c^2 \cdot U_{xx}$$

$$U(x, 0) = f(x) \quad for \quad 0 \le x \le L$$

$$U_x(0, t) = 0 = U_x(L, t) \quad for \quad t \ge 0$$

14.2. (The plucked string problem associated with the one-dimensional wave equation.) Given positive constants L, c and a differentiable function $f(x)$ on $[0, L]$, use separation of variables to find $U(x, t)$, defined for $0 \leq x \leq L$ and $t \geq 0$, such that

$$U_{tt} = c^2 \cdot U_{xx}$$
$$U(x, 0) = f(x) \quad \text{for} \quad 0 < x < L$$
$$U_t(x, 0) = 0 \quad \text{for} \quad 0 < x < L$$
$$U(0, t) = U(L, t) = 0 \quad \text{for} \quad t \geq 0$$

14.3. (A two-dimensional diffusion problem.) Let L, c be positive constants. Let $f(x, y)$ be defined on the square (x, y) with $0 \leq x \leq L$ and $0 \leq y \leq L$. Find $U(x, y, t)$, where $0 \leq x \leq L$ and $0 \leq y \leq L$, and $t \geq 0$ such that

$$U_t = c^2 \cdot \left[U_{xx} + U_{yy} \right]$$
$$U(x, y, 0) = f(x, y)$$
$$U(0, y, t) = U(L, y, t) = U(x, 0, t) = U(x, M, t) = 0$$

14.4. (Laplace's equation on a disk.[10]) We are given a differentiable and 2π-periodic function $f(\theta)$. Find $W(r, \theta)$, defined for $0 \leq r \leq 1$ and all θ such that

$$W(r, \theta + 2\pi) = W(r, \theta)$$
$$r^2 \cdot W_{rr} + r \cdot W_r + W_{\theta\theta} = 0$$
$$W(1, \theta) = f(\theta)$$

(Recall that the PDE for W is the polar version of Laplace's equation.)

---

[10]This is another *Dirichlet Problem*.

14.5. Let n be an integer. Show that $U_n(x, y) = (x + iy)^n + (x - iy)^n$ is harmonic[11] (that it satisfies Laplace's PDE). (Note: $i^2 = -1$, as usual. Assume you can use the usual differentiation rules even with i. You might like to compute U by multiplying out in the cases $n = 1, 2, -1$.)

14.6. (This problem shows that we cannot necessarily take the derivative of a Fourier series term by term, the way we do for Taylor series.) Find a sine series for the function x on $[0, 3]$. Take the derivative of the sine series term by term, and plug in $x = 3/2$; show that the resulting series cannot converge. (Hint: the Divergence Test.)

14.7. Find a cosine series for the function $|1 - x|$ on $[0, 2]$. (Hint: break up the interval of integration into $[0, 1]$ and $[1, 2]$.)

---

[11]If a_n is a sequence with radius of convergence r, then $U(x, y) = \sum_{n=0}^{\infty} a_n \cdot U_n(x, y)$ defines a harmonic function inside the circle $x^2 + y^2 = r^2$. Thus, every such sequence defines a solution to Laplace's equation. That's a lot of solutions!

Bibliography

[Standard DE textbooks] These days, textbooks come out in new editions every other year or so. To my mind, the changes year to year are not significant, so if you obtain any particular edition of either of these books, you will have an adequate reference.

[1] William E. Boyce and Richard C. DiPrima, *Elementary Differential Equations and Boundary Value Problems*, 8th edition, Hoboken, New Jersey: John Wiley & Sons, 2005.

[2] C.H. Edwards and David E. Penney, *Elementary Differential Equations*, 3rd edition, Englewood Cliffs, New Jersey: Prentice-Hall, 1985.

[Standard linear algebra textbooks] Books on this subject vary from computational to more theoretical. The two books listed here are reliable.

[3] Howard Anton, *Elementary Linear Algebra with Applications*, Hoboken, New Jersey: John Wiley & Sons.

[4] Bernard Kolman and David Hill, *Elementary Linear Algebra*, 7th ed., New Jersey: Prentice-Hall, 1995-2000.

[Sources for formulas and solution techniques]

[5] L. Euler, *Introduction to Analysis of the Infinite, Book I*, trans. John Blanton, New York: Springer, 1988.

[6] L. Euler, *On Differential Equations of the Second Order*, English translation in David Eugene Smith, *A Source Book in Mathematics*, New York: Dover Publications, 1959.

[7] W. Rogosinski, *Fourier Series*, Trans. Harvey Cohn, F. Steinhardt, New York: Chelsea Publishing Company, 1959.

[Historical information]

[8] David Brewster, Leonhard Euler, John Griscom,*Letters of Euler on Different Subjects in Natural Philosophy*, Volume 1, facsimile, New York: J & J Harper, 1833.

[9] H Lebesgue, L'oeuvre mathématique de Vandermonde, *Enseignement Math.* (2) 1 (1956), pp203-223.

[10] Willis I. Milham, *Time and Timekeepers*. New York: MacMillan, 1945.

[11] Thomas Muir, *A Treatise on the Theory of Determinants*, Dover Publishing, 2003.

[12] J.H. Poincaré, *Les Methods Nouvelles de al Mecanique Celeste*, vol. II, Paris: Gauthier-Villars.

[13] D.E. Smith, *History of Mathematics, Volume II*, Mineola, New York: Dover, 1958.

[References for application problems]

[14] Jay L. Devore and Kenneth N. Berk, *Modern Mathematical Statistics with Applications*, Belmont, California: Thomson Brooks-Cole, 2007.

[15] Robert Horton, Laurence A. Moran, Gray Scrimgeour, Marc Perry, David Rawn, *Principles of Biochemistry*, 4th-edition, Prentice Hall, 2006.

[16] George Shortley and Dudley Williams, *Elements of Physics*, 5th edition, Englewood Cliffs, New Jersey: Prentice-Hall, 1971.

Index